T0230431

BestMasters

Springer awards „BestMasters" to the best master's theses which have been completed at renowned universities in Germany, Austria, and Switzerland.

The studies received highest marks and were recommended for publication by supervisors. They address current issues from various fields of research in natural sciences, psychology, technology, and economics.

The series addresses practitioners as well as scientists and, in particular, offers guidance for early stage researchers.

Carolin Loos

Analysis of Single-Cell Data

ODE Constrained Mixture
Modeling and Approximate
Bayesian Computation

 Springer Spektrum

Carolin Loos
München, Germany

BestMasters
ISBN 978-3-658-13233-0   ISBN 978-3-658-13234-7 (eBook)
DOI 10.1007/978-3-658-13234-7

Library of Congress Control Number: 2016935216

Springer Spektrum

Printed on acid-free paper

This Springer Spektrum imprint is published by Springer Nature
The registered company is Springer Fachmedien Wiesbaden GmbH

Acknowledgements

I would like to begin by offering my sincerest gratitude to my supervisor Dr. Jan Hasenauer for his immense support, advice and encouragement, and for truly inspiring my interest in this field of research. I also gratefully acknowledge Dr. Carsten Marr for always taking the time to answer my questions and help me with my problems. Furthermore, I would like to thank the members of the groups Data-driven Computational Modeling and Quantitative Single Cell Dynamics for their feedback, explanations and interesting discussions. I am highly indebted to Prof. Dr. Dr. Fabian Theis for giving me the opportunity to work on this interesting project at the ICB. It was a pleasure to explore this fascinating field of research and to write my thesis in such an inspiring working environment. Last, but not least, I would like to thank my family and friends for their company and constant support throughout the years.

Carolin Loos

Abstract

Investigating cellular heterogeneity is of great importance for a holistic understanding of biological processes and is therefore a focus of systems biology. This task requires sophisticated models of single-cell data, which in turn need parameter estimation approaches that are able to fit these models to given measurement data.

The first part of this thesis focuses on using ODE constrained mixture models (ODE-MMs) for the analysis of single-cell snapshot data. With these models subpopulations can be identified and even the source of differences between subpopulations can be detected. We investigate the method's applicability to the study of the alteration of subpopulation response by the cellular environment with novel data of NGF-induced Erk signaling, a process relevant in pain sensitization. We enhance the method by providing a mechanistic description of the variability of the subpopulations using moment equations. In addition, we propose ODE-MMs for the analysis of multivariate measurements, which accounts for correlations among the measurands. Applying our method to artificial data of a conversion process and to real multivariate data for NGF-induced phosphorylation of Erk enables an improved insight into the underlying system.

In the second part of this thesis, we study stochastic dynamics of individuals cells that are modeled with continuous time Markov chains (CTMCs). We introduce a likelihood-free approximate Bayesian computation (ABC) approach for single-cell time-lapse data. This method uses multivariate statistics on the distribution of single-cell trajectories. We evaluate our method for samples of a bivariate normal distribution and for artificial equilibrium and non-equilibrium single-cell time-series of a one-stage model of gene expression. In addition, we assess our method by applying it to data generated with parameter variability and to tree-structured time-series data. A comparison with an existing method using statistics reveals an improved parameter identifiability using multivariate statistics.

In summary, this thesis introduces two novel approaches for the analysis of multivariate data that can be used to study cellular heterogeneity based on single-cell data.

Kurzfassung

Ein tiefgehendes Verständnis für die Mechanismen von biologischen Prozessen erfordert die Erforschung der Heterogeneität von Zellpopulationen. Aus diesem Grund bildet die Untersuchung heterogener Zellpopulationen aktuell einen Forschungsschwerpunkt in der Systembiologie. Hierfür werden komplexe mathematische Modelle benötigt, welche wiederum Parameterschätzungsmethoden erfordern, die in der Lage, sind diese Modelle mit gegebenen Messdaten zusammenzuführen.

Der erste Teil dieser Arbeit befasst sich mit der Analyse von Einzelzelldaten, für welche jeweils die Verteilungen der gemessenen Konzentrationen in den Zellen zu einem bestimmten Zeitpunkt gegeben sind. Für diese Analyse nutzen wir *ODE constrained mixture models* (ODE-MMs), sogenannte Mischmodelle, welche durch gewöhnliche Differentialgleichungen beschränkt sind. Mit diesen Modellen können Subpopulationen innerhalb einer Zellpopulation ermittelt, und sogar die Ursache für den Unterschied zwischen den Subpopulationen identifiziert werden. Wir verwenden diese Methode erstmals zur Untersuchung von Veränderung von Subpopulationsreaktionen aufgrund der Zellumgebung. Diese Analyse erfolgt auf neuen Daten für die durch NGF induzierte Phosphorylierung von Erk, ein für die Schmerzsensitivierung relevanter Prozess. Wir verbessern die Methode, indem wir Momentengleichungen für die mechanistische Beschreibung der Subpopulationen verwenden. Darüber hinaus entwickeln wir ODE-MMs zur Analyse von multivariaten Messungen, wodurch Korrelationen zwischen Messungen berücksichtigt werden. Wir testen unsere Methode anhand von artifiziellen Daten eines Konversionsprozesses und anhand multivariater Messungen für NGF induzierte Erk-Phosphorylierung. Es zeigt sich, dass unsere Methode einen genaueren Einblick in das zugrundeliegende biologische System ermöglicht.

Im zweiten Teil dieser Arbeit analysieren wir stochastische Dynamiken von individuellen Zellen, welche durch Markovketten in stetiger Zeit modelliert werden. Hierfür stellen wir eine auf *Approximate Bayesian Computation* (ABC) basierende, Likelihood-freie Methode zur Parameterschätzung von Einzelzellzeitreihen vor. Diese Methode nutzt multivariate Statistiken auf der Verteilung von Einzelzelltrajektorien. Wir evaluieren unsere Methode sowohl für Daten, die durch eine bivariate Normalverteilung generiert wurden als auch für artifizielle Einzellzellzeitreihen eines einstufigen Genexpressionmodells, welche sich in und außerhalb ihres stationären Gleichgewichts befinden. Der Vergleich mit einer existierenden Method, die Statistiken verwendet, verdeutlicht, dass durch eine multivariate Betrachtung die Modellparameter besser identifiziert werden können.

Zusammenfassend werden in dieser Arbeit zwei neuartige Methoden zur Analyse von multivariaten Daten entwickelt. Diese sind geeignet, um heterogene Zellpopulationen basierend auf Einzelzelldaten zu untersuchen.

Contents

List of Figures

List of Tables

List of Abbreviations

ABC	approximate Bayesian computation
ABC SMC	approximate Bayesian Computation with sequential Monte Carlo
AIC	Akaike information criterion
BIC	Bayesian information criterion
CM	cross-match test
CME	chemical master equation
CTMC	continuous time Markov chain
FSP	finite state projection
KS	Kolmorgorov-Smirnov
MAP	maximum a posteriori
MCMC	Markov chain Monte Carlo
ME	moment equation
MLE	maximum likelihood estimate
MME	maximum mean discrepancy
ODE	ordinary differential equation
ODE-MM	ODE constrained mixture model
RRE	reaction rate equation
SSA	stochastic simulation algorithm

List of Symbols

Chapter 2

$\boldsymbol{\theta}$ parameter vector

$\mathcal{D}$ data

ν_{ij} stochiometric coefficient

$a_j(\mathbf{x})$ propensity function for reaction j

$L(\boldsymbol{\theta})$ likelihood function of $\boldsymbol{\theta}$ given $\mathcal{D}$ (conditional probability)

$p(\boldsymbol{\theta})$ prior probability of $\boldsymbol{\theta}$

$p(\boldsymbol{\theta}|\mathcal{D})$ posterior probability of $\boldsymbol{\theta}$ given $\mathcal{D}$

Chapter 3

$\boldsymbol{\psi}$ parameters of a subpopulation

$\log \mathcal{N}(\mathbf{y}|\boldsymbol{\mu}, \boldsymbol{\Sigma})$ multivariate log-normal distribution

$\log \mathcal{N}(y|\mu, \sigma^2)$ univariate log-normal distribution

$\mathcal{N}(\mathbf{y}|\boldsymbol{\mu}, \boldsymbol{\Sigma})$ multivariate normal distribution

$\mathcal{N}(y|\mu, \sigma^2)$ univariate normal distribution

$\mathbf{C}$ covariances of the species of the system

$\mathbf{m}$ means of the species of the system

$\mathbf{x}$ state vector of the moments of the species

$C_y, \mathbf{C_y}$ (co)variance(s) of the measurand(s)

f function describing time evolution of the moments

h	linking function of mixture parameters and state vector of the moments
$m_y, \mathbf{m_y}$	mean(s) of the measurand(s)
n_s	number of subpopulations
$p(y\|\boldsymbol{\varphi})$	mixture probability
u	external stimulus
w_s^e	weight of subpopulation s under experiment e

Chapter 4

α	confidence level
ϵ_t	threshold for population t
ϵ_{end}	threshold for final population
γ	mRNA degradation rate
$\hat{F}_{\mathbf{X}}$	empirical cumulative distribution
λ	mRNA synthesis rate
$\mathcal{D}^{\text{obs}}$	observed data
$\mathcal{D}^{\text{sim}}$	simulated data
$\mathcal{S}(\mathcal{D})$	summary statistic of data $\mathcal{D}$
$\mathbf{X}$	samples of $\mathcal{D}^{obs}$
$\mathbf{Y}$	samples of $\mathcal{D}^{sim}$
A_l	random variable counting pairs with exactly l nodes of observed samples
c_{max}	number of cross-matches for final population
$d_{\text{KS}}(\cdot, \cdot)$	Kolmogorov-Smirnov distance
$k(\cdot, \cdot)$	Gaussian kernel

$K_t(\cdot|\cdot)$	perturbation kernel
m	number of simulated data points
n	number of observed data points
n_t	number of measured time points
P	number of particles per populations
$p(\mathcal{D}^{sim}|\boldsymbol{\theta})$	simulation function
p, q	underlying distributions of $\mathbf{X}$ and $\mathbf{Y}$
T	number of populations
$w_i^{(t)}$	weight of particle $\theta_i^{(t)}$ in population t

1 Introduction

The goal of systems biology is to understand biochemical processes as a whole (Kitano, 2002). This is accomplished by analyzing biological experimental data with computational models that describe the dynamical behavior of the system (Cho & Wolkenhauer, 2005). Often population averaged data is considered, which only contains information about the mean behavior of the cells. Such data is for example produced by microarrays (Malone & Oliver, 2011) or Western blots (Renart et al., 1979). Using population averaged data, cellular heterogeneity, i.e., differences among isogenic cells, can not be captured and subpopulation structures remain concealed (Altschuler & Wu, 2010).

Eludicating heterogeneity is a goal of current research, as it has been shown to have important implications for cell fate decisions. The consequences of heterogeneity have been studied for several types of cells, ranging from stem cells (e.g. (Torres-Padilla & Chambers, 2014)) to cancer cells (e.g. (Michor & Polyak, 2010)).

1.1 Modeling and Parameter Estimation for Single-Cell Data

Analysis of heterogeneity requires measurements performed at the single-cell level, which can be obtained using techniques such as flow cytometry (Pyne et al., 2009) or fluorescent microscopy (Muzzey & Oudenaarden, 2009; Schroeder, 2011; Miyashiro & Goulian, 2007). Elowitz et al. (2002) placed two identically regulated reporter genes in the same cell and identified different sources of heterogeneity by analyzing the corresponding single-cell data. The overall variation of gene expression can be partitioned into extrinsic and intrinsic noise. Variability that affects both reporter genes equally corresponds to extrinsic noise. Differences in gene expression arising due to random births and deaths of single molecules are called intrinsic noise.

While most models only describe the mean behavior of the cells (Resat et al., 2009), studying heterogeneity requires more detailed models, which take the single-cell nature of the data into account. A possibility to incorporate intrinsic noise is to model birth and death processes of individual molecules as continuous time Markov chains (CTMCs) with stochastic chemical kinetics (Gillespie, 2007). There are also deterministic approaches, which describe statistics of CTMCs and account for variability instead of only considering the mean behavior. An example of such an approach is the method of moments (Engblom, 2006).

Appropriate models for experimental data should not only capture important properties of the system that are being investigated but should also consider the trade-off between simplicity and accuracy. When the dimension of the parameter space increases, the model generally loses predictive power. Moreover, when the model gets more detailed the simulation of the model gets more complex (Wilkinson, 2009).

Understanding heterogeneity requires efficient parameter inference, since studying a data-based model requires knowledge of the model parameters such as kinetic rates and initial conditions. However, most of these parameters can not be measured experimentally and need to be estimated from the available data (Lillacci & Khammash, 2010). While standard approaches to estimate parameters from observed data maximize the likelihood function, a function that represents the probability of observing a data set given some parameters, this approach is intractable for many stochastic models as the likelihood function is computationally too costly. This problem is tackled by using likelihood-free methods, which are also called approximate Bayesian computation (ABC) methods (Marjoram *et al.*, 2003). These methods circumvent the evaluation of the likelihood function by comparing observed and simulated data sets. Unfortunately, inferring the model parameters based on experimental data in general gets more challenging for stochastic models (Wilkinson, 2009).

1.2 Contribution of this Thesis

In this thesis, we study computational models that account for heterogeneity in cell populations. We calibrate these models to artificial and real experimental data at the single-cell level with parameter estimation techniques that are suited to the complexity of the models. This thesis is structured as follows:

Chapter 2 introduces two types of single-cell data and presents the key concepts needed for their analysis. This comprises computational modeling of the data, which uses stochastic chemical kinetics. Experimental data and the derived models are fitted by performing parameter inference.

In Chapter 3, we focus on ODE constrained mixture modeling (ODE-MM), an approach that combines mixture probabilities and a mechanistic description for the behavior of individual subpopulations of a cell population. We evaluate the method of ODE-MMs by using it for the detection of altered subpopulation responses under different experimental conditions. For this, we consider novel single-cell snapshot data of NGF-induced Erk phosphorylation. In addition, we enhance the method by using moment equations for the description of the underlying biological process. Moreover, in order to gain even more information from the data we develop the method for the analysis of multivariate measurements. We assess our method by applying it to artificial data of a conversion process and to real experimental data of NGF-induced phosporylation of Erk.

In Chapter 4, we develop an ABC method using multivariate test statistics for single-cell time-lapse data that are modeled with CTMCs. We introduce two multivariate test statistics and evaluate the respective ABC methods on a bivariate normal distribution. In addition, we apply our method to artificial single-cell time-lapse data of a one-stage model of gene expression, accounting for extrinsic cell-to-cell variability and cell division.

In Chapter 5 we summarize our results and draw conclusions.

2 Background

This chapter introduces the key concepts that are needed to understand this thesis. First, we describe the different types of experimental data that are analyzed. Afterwards, the principles of modeling of chemical kinetics are introduced with a focus on the chemical master equation (CME) and its approximations. Finally, we show how experimental data and biological models can be brought together with inference. Inference consists of parameter optimization, identifiability and uncertainty analysis, and model selection.

2.1 Experimental Data

In this thesis, we consider and distinguish two different types of single-cell data $\mathcal{D}$ that provide information about cell-to-cell variability and are frequently collected in biological research.

Single-cell snapshot data $\mathcal{D} = \{\{y_j(t_k)\}_j\}_{k=1}^{n_t}$ provide single-cell measurements for n_t time instances t_k (see Figure 2.1A). Common approaches to generate these data are, e.g., flow cytometry (Davey & Kell, 1996) or single-cell microscopy (Miyashiro & Goulian, 2007). A key advantage of these technologies is the possibility of measuring many genes of plenty of single-cells with low costs. As the cells are not tracked over time, no information about the time-course of an individual cell is available.

To obtain temporal information single-cell time-lapse data $\mathcal{D} = \{\{y_j(t_k)\}_{k=1}^{n_t}\}_j$ (see Figure 2.1B) are required. Single-cell time-lapse data are typically obtained by conducting fluorescent time-lapse microscopy (Muzzey & Oudenaarden, 2009) followed by single-cell tracking (Schroeder, 2011) and image analysis. This approach provides a smaller number of cells than the technologies described before and the generation of single-cell time series is expensive and time-consuming. On the other hand, cells are tracked over time yielding a higher information content of the data.

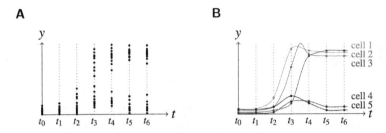

Figure 2.1: Measurement data at the single-cell level adopted from Hasenauer (2013): **(A)** Illustration of single-cell snapshot data of some measurement y. **(B)** Illustration of single-cell time-lapse data for five individual cells.

2.2 Modeling Chemical Kinetics

For the detailed analysis of single-cell data, mechanistic mathematical models are used. One possibility is the use of stochastic chemical kinetics, which model biochemical reaction networks as continuous-time discrete-state Markov chains (CTMCs). The time evolution of a CTMC is governed by the CME. A process defined by the CME can either be simulated with the stochastic simulation algorithm (SSA) or its solution can be approximated e.g. with the moment equations (ME). While stochastic modeling is especially important in the case of low-copy numbers, we assume that for high numbers of molecules the system can be described by its average behavior. This can be modeled in a deterministic way by first order ordinary differential equations (ODEs) describing the evolution of concentrations of the species.

2.2.1 Stochastic Chemical Kinetics

Stochastic models are mostly used to describe a biological process, when it is important to consider that molecules only appear in whole numbers (Wilkinson, 2009; Resat *et al.*, 2009). This discreteness yields a stochasticity in the dynamics of the molecules and especially has to be taken into account if only few numbers of molecules are present.

Stochastic chemical kinetics describe the time evolution of a chemical system consisting of L chemical species $x_1, \ldots, x_L$ that interact inside a volume Ω through M reactions $R_1, \ldots, R_M$. A reaction R_j has the form

$$\nu_{1j}^- x_1 + \ldots + \nu_{Lj}^- x_L \xrightarrow{k_j} \nu_{1j}^+ c_1 + \ldots + \nu_{Lj}^+ x_L,$$

with stochiometric coefficients $\nu_{ij}^+, \nu_{ij}^- \in \mathbb{N}_0$ and reaction rate k_j. A state of the system is represented by a vector $\mathbf{x}(t) \in \mathbb{N}_0^L$. Each entry of the vector is the number of molecules of the corresponding species. The stochiometric matrix $\mathbf{S} = (\mathbf{s}_1, \ldots, \mathbf{s}_M) \in \mathbb{R}^{L \times M}$ is defined by $\{S_{ij}\} = \{\nu_{ij}^+ - \nu_{ij}^-\} := \{\nu_{ij}\}$. Each entry of the matrix describes the change in the number of molecules of species x_i due to a reaction of type j, i.e., the state $\mathbf{x}$ changes to $\mathbf{x} + \mathbf{s}_j$ after reaction R_j took place. The probability that reaction R_j happens in the next infinitesimal time interval $[t, t + dt)$ is $a_j(\mathbf{x})dt$, with propensity function $a_j(\mathbf{x})$.

Several assumptions are typically made when deriving a model of a biological process, e.g. that the system has a constant volume Ω and is well-stirred, i.e., the probability of some molecules of a species being in one particular region is uniform over the volume (Gillespie, 2007). We consider zero-order reactions, which are independent of the number of molecules, unimolecular reactions, in which just a single molecule is necessary to conduct the reaction, and bimolecular reactions, for which two molecules need to collide. Higher order reactions can easily be integrated into the methods proposed in this thesis.

2.2.2 Chemical Master Equation

The CME governs the evolution of the probability that the stochastic process is in a particular state, given by $p(\mathbf{x}, t)$, over time (Gillespie, 1992). The probability $p(\mathbf{x}, t | \mathbf{x}_0, t_0)$ is conditioned on the system being in state $\mathbf{x}_0$ at time t_0. To obtain an evolution equation the probability $p(\mathbf{x}, t + dt | \mathbf{x}_0, t_0)$ is first derived in terms of $p(\mathbf{x}, t | \mathbf{x}_0, t_0)$, by assuming dt is small enough that at most one reaction can occur in the time interval $[t, t + dt)$. One possibility for the system being in state $\mathbf{x}$ at time $t + dt$ is that it already has been in this state and no reaction has taken place since time t, which happens with probability $1 - \sum_{j=1}^{M} a_j(\mathbf{x})dt + \mathcal{O}(dt)$. Another scenario is that the system has been in state $\mathbf{x} - \mathbf{s}_j$ and a reaction of type j occurred with probability $a_j(\mathbf{x} - \mathbf{s}_j)dt$, which yields M more possibilities. After summing up the probabilities and taking the limit $dt \to 0$, we obtain the CME

$$\frac{dp(\mathbf{x}, t | \mathbf{x}_0, t_0)}{dt} = \sum_{j=1}^{M} \left[p(\mathbf{x} - \mathbf{s}_j, t | \mathbf{x}_0, t_0) a_j(\mathbf{x} - \mathbf{s}_j) - a_j(\mathbf{x}) p(\mathbf{x}, t | \mathbf{x}_0, t_0) \right],$$

with initial condition

$$p(\mathbf{x}, t = t_0 | \mathbf{x}_0, t_0) = \begin{cases} 1, & \mathbf{x} = \mathbf{x}_0 \\ 0, & \mathbf{x} \neq \mathbf{x}_0 \end{cases}.$$

If we neglect $\mathbf{x}_0$ and t_0 for a simpler notation we obtain

$$\frac{dp(\mathbf{x}, t)}{dt} = \sum_{j=1}^{M} \left[p(\mathbf{x} - \mathbf{s}_j, t) a_j(\mathbf{x} - \mathbf{s}_j) - a_j(\mathbf{x}) p(\mathbf{x}, t) \right],$$

with initial condition $p(\mathbf{x}, t_0) = p_0(\mathbf{x})$. The CME indeed completely determines the probability $p(\mathbf{x}, t | \mathbf{x}_0, t_0)$ and thus totally describes the system. However, it consists of a system of coupled ordinary differential equations (ODEs), with one ODE for every possible state of the system. Since the state space of a biological system is mostly high dimensional or even infinite dimensional, the CME can only be solved analytically or in a feasible numerical way for a few simple cases (e.g. (Jahnke & Huisinga, 2007)).

2.2.3 Stochastic Simulation Algorithm

Instead of solving the CME, it is possible to simulate samples in form of trajectories and thereby recover the underlying probability distribution. This is motivated by the fact that the chance of a particular trajectory being simulated corresponds to the probability given by the CME. A possibility to obtain trajectories is the SSA (Gillespie, 1977). This algorithm enables an exact simulation of trajectories consistent with the probability distribution and the transition probabilities that are associated with the CME. For the direct method of stochastic simulation we define

- the sum over all reaction propensities $a_0(\mathbf{x}) = \sum_{j=1}^{M} a_j(\mathbf{x})$,
- the time τ to the next reaction,
- the index j of the next reaction.

It can be shown that τ is exponentially distributed with rate $a_0(\mathbf{x})$ and j has density $\frac{a_j(\mathbf{x})}{a_0(\mathbf{x})}$, which yields the following algorithm:

Algorithm 2.1: Direct method

Input: Initial condition $\mathbf{x}_0 \in \mathbb{N}_0^L$,

final simulation time t_{end},

reaction propensity functions $a_j(x), j = 1, \ldots, M$,

stochiometric matrix $\mathbf{S} = (\mathbf{s}_1, \ldots, \mathbf{s}_M) \in \mathbb{Z}^{L x M}$.

Result: Time trajectory of state vector $\mathbf{x}(t)$.

Set $t \leftarrow 0$ and $\mathbf{x} \leftarrow \mathbf{x}_0$.

while $t < t_{\text{end}}$ **do**

Evaluate reaction propensity functions $a_j(\mathbf{x})$ and calculate $a_0(\mathbf{x}) = \sum_{j=1}^M a_j(x)$.

Generate two uniformly distributed independent random numbers r_1 and r_2.

Calculate the time until the next reaction takes places by $\tau = \frac{1}{a_0(\mathbf{x})} \log(1/r_1)$.

Find the index j of the next reaction that satisfies $\sum_{j=1}^M a_j(\mathbf{x}) > r_2 a_0(x)$.

Update the state of the system $\mathbf{x} \leftarrow \mathbf{x} + \mathbf{s}_j$.

Update the time $t \leftarrow t + \tau$.

end

An example of trajectories obtained by this method is shown in Figure 2.2 for a conversion process (see Section 3.3.2). The computation can be inefficient if lots of events have to be simulated. Therefore, approximations such as τ-leaping have been introduced (for further information see (Gillespie, 2007)).

2.2.4 Method of Moments

A possibility to approximate the solution of the CME and thereby avoid the computational costs of the SSA is the method of moments (Engblom, 2006). This method computes the moments of $p(\mathbf{x}, t)$, i.e., the mean

$$m_i(t) = \sum_{x \in \Omega} x_i p(\mathbf{x}, t), \quad i = 1, \ldots, L,$$

of species x_i, and higher order moments such as the covariance

$$C_{ij}(t) = \sum_{x \in \Omega} (x_i - m_i(t))(x_j - m_j(t)) p(\mathbf{x}, t), \quad i, j = 1, \ldots, L,$$

of species x_i and x_j. The time evolution of the moments is described by a set of ODEs, the so-called moment equations (MEs). If the system comprises bimolecular reactions, the calculation of higher order moments is recursive, i.e., the evolution of a moment of order k depends on moments of order $k + 1$. In this case moment closure techniques must be applied, introducing an approximation error (Lee *et al.*, 2009). Formulas for the first and second order moments of system with at most bimolecular reactions, can be found in (Engblom, 2006, Proposition 2.5.). The first and second order moments, namely mean and variance, of the solution statistics for a conversion process are depicted in Figure 2.2. If a system comprises low- and medium/high-copy species the method of conditional moments (Hasenauer *et al.*, 2014a) can be used. This method conditions the moments of species with medium or higher abundance on the states of species that are only present in low-copy numbers. Therefore, it accounts for the stochasticity of the processes, arising due to the discreteness of the low-abundance species. The method avoids the computational costs arising from a full stochastic description of the system using MEs for the medium and high-copy species.

2.2.5 Reaction Rate Equation

In the limit of large numbers of molecules, the system behaves in a more deterministic way and the importance of considering single molecules vanishes. Therefore, measurements are at a continuous level, in contrast to the discrete state space of stochastic modeling. The evolution of the system is captured by the reaction rate equations (RREs) (Resat *et al.*, 2009; Gillespie, 2007)

$$\frac{d\mathbf{x}(t)}{dt} = \sum_{j=1}^{M} \mathbf{s}_j a_j(\mathbf{x}(t)) \, .$$

For some simple systems an explicit formula for the solution of the RREs can be derived, but mostly numerical integration is need. Nevertheless, deterministic simulations of a system are generally faster than a stochastic simulation (Szekely & Burrage, 2014).

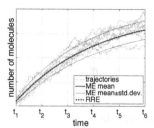

Figure 2.2: Example of trajectories of one species of a conversion process obtained by the SSA (gray), the corresponding approximation with MEs (red) and RREs (blue), where the mean described by ME and the RRE coincide.

2.3 Parameter Inference

The idea of parameter inference is to combine observed data $\mathcal{D}$ and a model $\mathcal{M}$, which for example has been derived with techniques presented in the previous section. Such a model comprises parameters, for example kinetic rates or initial conditions, and some of these parameters denoted by $\boldsymbol{\theta} \in \mathbb{R}^{n_\theta}$ may be unknown, because either they are not measured or it is impossible to measure them.

2.3.1 Parameter Estimation

A common approach to estimate the parameters of a model is to maximize the likelihood function

$$L(\boldsymbol{\theta}) = p(\mathcal{D}|\boldsymbol{\theta}),$$

which describes the conditional probability of observing $\mathcal{D}$ given $\boldsymbol{\theta}$. Due to better numerical properties for optimization, usually the negative log-likelihood function

$$J(\boldsymbol{\theta}) = -\log L(\boldsymbol{\theta})$$

is minimized. The parameters $\boldsymbol{\theta}^{\mathrm{ML}}$ that maximize the likelihood function (or minimize the negative (log-)likelihood function) are called the maximum likelihood estimates (MLE).

In a Bayesian framework we can additionally incorporate prior knowledge about the parameters using the prior distribution $p(\boldsymbol{\theta})$ (Hastie *et al.*, 2009). Applying Bayes' theorem yields the posterior distribution of the parameters

$$p(\boldsymbol{\theta}|\mathcal{D}) = \frac{p(\mathcal{D}|\boldsymbol{\theta})p(\boldsymbol{\theta})}{p(\mathcal{D})} \propto p(\mathcal{D}|\boldsymbol{\theta})p(\boldsymbol{\theta}) \,.$$

The parameters $\boldsymbol{\theta}^{\mathrm{MAP}}$ that maximize the posterior distribution are the maximum a posteriori estimate (MAP), the Bayesian counterpart of the MLE. The evaluation of the normalizing constant $p(\mathcal{D}) = \int p(\mathcal{D}|\boldsymbol{\theta})p(\boldsymbol{\theta})d\boldsymbol{\theta}$ can be computationally expensive or unfeasible. However, this constant can be neglected for optimization and uncertainty analysis, as it is only needed for model selection based on Bayes factors (Raftery, 1999). The minimization of the negative log-likelihood function can be efficiently performed using multi-start local optimization. For this, the initial values for the optimizer are e.g. obtained by Latin hypercube sampling and then are chosen in a sequential way, such that the corresponding objective function values are decreasing (Raue *et al.*, 2013). For the optimization procedure the calculation of the gradient is of great importance, as the derivative of the objective function is used to determine the next parameter value. For the calculation of the derivatives finite differences or sensitivity analysis can be used (Sengupta *et al.*, 2014). Sensitivity analysis describes the derivatives of the objective function with respect to the parameters. Using them, the gradient can be calculated numerically more robustly. Additionally, we use log-transformed parameters $\boldsymbol{\xi} = \log(\boldsymbol{\theta})$ due to better convergence properties.

If the likelihood cannot be expressed analytically or is computationally too costly to evaluate, so-called likelihood-free parameter estimation methods are required. This class of methods circumvents the calculation of the likelihood function and is also known under the name approximate Bayesian computing (ABC) (Csilléry *et al.*, 2010). We explain these methods in more detail in Section 4.2, as they are the focal point of the work described in Chapter 4.

2.3.2 Identifiability and Uncertainty Analysis

Due to the structure of the examined system and limitations of the available data some parameters can be non-identifiable (Raue *et al.*, 2009), i.e., the parameter can not be determined from the data. If this is the case even for perfect data, the parameter is structurally non-identifiable. If the parameter can not be identified due to measurement noise or too little data, the parameter is practically non-identifiable. Studying these uncertainties is an important step of parameter inference and explained in the following.

A common approach to analyze uncertainties of the parameters is to calculate confidence intervals, e.g. asymptotic confidence intervals based on the curvature of the likelihood, such as the hessian, or finite sample confidence intervals based on profile likelihoods (for further information see (Raue *et al.*, 2009)). A parameter θ is practically identifiable from the corresponding data, if the corresponding confidence intervals are finite.

In a Bayesian context, in which parameters are treated as random variables, we can get information about the uncertainty of the estimates by considering the whole posterior distribution. Because of a possibly high dimension of the parameter space or the lack of a closed form for the posterior, the use of numerical sampling from the posterior distribution is required. Samples from the posterior distribution can be obtained by Markov chain Monte Carlo (MCMC) methods (Gilks *et al.*, 1996).

2.3.3 Model Selection

The last step of parameter inference is to select an optimal model of out a given set of candidate models $\mathcal{M}_1, \ldots, \mathcal{M}_l$ that have been derived for some data $\mathcal{D}$. On the one hand, the chosen model should fit the data very well, which can be easily improved by increasing the number of parameters. On the other hand, the model should be as simple as possible to provide reliable predictions and avoid unnecessary uncertainties. We introduce two existing criteria for model selection that try to solve the trade-off between over- and underfitting of the data. Both criteria consist of a term with the likelihood value of the maximum likelihood estimate and a penalization term for a higher complexity of the model.

The Akaike information criterion (AIC) is based on information theoretical concepts (Akaike, 1998). It gives an estimate for Kullback-Leibler divergence between the densities of the true unknown model and of a candidate model $\mathcal{M}_k$ by

$$\mathrm{AIC}_k = -2\log(p(\mathcal{D}|\boldsymbol{\theta}^{\mathrm{ML},k})) + 2n_{\theta,k}\,,$$

with $\boldsymbol{\theta}^{\mathrm{ML},k}$ denoting the MLE for model $\mathcal{M}_k$ and $n_{\theta,k}$ denoting the number of parameters of the model. A low value of the AIC indicates that less information has been lost considering the candidate model and therefore a higher reliability. We reject models with $\Delta_{\mathrm{AIC}} = \mathrm{AIC}_k - \mathrm{AIC}_{\mathrm{min}} > 10$ as proposed by Burnham & Anderson (2002).

A Bayesian criterion for model selection can be derived by examining the posterior probability $p(k|\mathcal{D})$ of model $\mathcal{M}_k$ (see (Schwarz et al., 1978) for further information). This criterion is called the Bayesian information criterion (BIC),

$$\mathrm{BIC}_k = -2\log(p(\mathcal{D}|\boldsymbol{\theta}^{\mathrm{ML},k}) + \log(n_{\mathcal{D}})n_{\theta,k}\,,$$

with $n_{\mathcal{D}}$ denoting the number of data points. As with the AIC, the model with the lowest BIC is chosen and we reject models with $\Delta_{\mathrm{BIC}} = \mathrm{BIC}_k - \mathrm{BIC}_{\mathrm{min}} > 10$ (Raftery, 1999).

In summary, this chapter outlined the key principles that are used in the following chapters of this thesis. We introduced single-cell snapshot data and single cell time-lapse data, which possess different information contents and number of data points. We discussed different approaches to solve the CME, ranging from exact solutions obtained with the SSA to approximations with MEs and showed the link to deterministic modeling by RREs. Moreover, this chapter contains an introduction to parameter inference, including parameter estimation, identifiability and uncertainty analysis, and model selection. We presented the approach of maximum likelihood estimation using multi-start local optimization, and defined the posterior distribution that is used in a Bayesian context. For identifiability and uncertainty analysis, profile likelihoods and MCMC sampling schemes can be used. Finally, we introduced the AIC and BIC, two criteria used for model selection.

3 ODE Constrained Mixture Modeling for Multivariate Data Using Moment Equations

The focus of this chapter is to assess, improve and extend ODE constrained mixture modeling (ODE-MM) (Hasenauer *et al.*, 2014b), a method for studying dynamics and structures
of subpopulations. In Section 3.1, we introduce the underlying method and formulate the
problems that are subsequently addressed in the following sections. In Section 3.2, we
apply ODE-MMs to novel single-cell snapshot data for NGF-induced Erk signaling. We
evaluate the applicability of ODE-MMs to unravel alteration of subpopulation response
by cellular environment. In addition, we increase the insight into the underlying biological
system that can be gained using ODE-MMs. In Section 3.3, we present ODE-MMs with
moment equations (Engblom, 2006) for the mechanistic description of a biological process, which yields the ability to account for variability within a subpopulation. Additional
knowledge of the system can also be gained by considering multivariate measurements simultaneously. For this, Section 3.4 provides a likelihood for ODE-MMs that is able to
take correlations between the measurements into account. Our method is validated for
the example of a conversion process and applied to real experimental multivariate data
of NGF-induced Erk signaling (Section 3.5). The results are summarized in Section 3.6,
in which we outline possible further extensions and improvements of our method.

3.1 Introduction and Problem Statement

Cell populations exhibit different degrees of heterogeneity caused by cell-to-cell variability.
Even cells with similar cell types can respond differently to identically stimuli. Studying
cell heterogeneity and its sources is important for a holistic understanding of the underlying biological processes and cellular mechanisms. Therefore, this task comprises not only
the identification of subpopulations, but also the detection of how the subpopulations
differ.

3.1.1 ODE Constrained Mixture Modeling

A recently presented method using ODE constrained mixture models (ODE-MM) (Hasenauer *et al.*, 2014b) can describe the mechanisms of a process and at the same time is able to exploit subpopulation structures. This is achieved by modeling subpopulation dynamics with RREs and treating different subpopulations as individual components of a mixture distribution. Combining these two approaches, the method benefits from both, the possibility to include distributional information and from getting mechanistic insights using ODEs. Using ODE-MMs, population snapshot data can be analyzed across different experimental conditions. Moreover, it has been shown that even the causal differences between subpopulations can be revealed.

Based on single-cell snapshot data $\{\mathcal{D}_k^e\}_{e,k}$ the unknown parameters $\boldsymbol{\theta} = \{(w_s^e, \boldsymbol{\psi}_s^e, \sigma_s^e)\}_{s,e}$ of properties of the n_s subpopulations can be estimated maximizing the likelihood function

$$L(\boldsymbol{\theta}) := \prod_{e,k,j} \sum_{s=1}^{n_s} w_s^e p\left(y_j^{e,k} | \boldsymbol{\varphi}_s^e(t_k)\right) \tag{3.1}$$

$$\text{s.t. } \dot{\mathbf{x}}_s^e = f\left(\mathbf{x}_s^e, \boldsymbol{\psi}_s^e, u^e\right), \ \mathbf{x}_s^e(0) = \mathbf{x}_0(\boldsymbol{\psi}_s^e, u^e),$$

$$\boldsymbol{\varphi}_s^e = h(\mathbf{x}_s^e, \boldsymbol{\psi}_s^e, u^e).$$

The indices e, k and j are for the experimental conditions, the time point and the single-cells, respectively. Moreover, $p\left(y_j^{k,e} | \boldsymbol{\varphi}^e(t_k)\right)$ is a mixture distribution, e.g. a normal or log-normal distribution with mixture parameters $\boldsymbol{\varphi}_s^e = (\mu_s^e, \sigma_s^e)$ and mixture weights w_s^e that sum up to 1. The ODE model given by reaction rate equations (RREs) is denoted by f. The means are linked to the RRE model by the function h. The variances needs to be estimated from the data and therefore are listed in the parameter vector $\boldsymbol{\theta}$. The parameters $\boldsymbol{\psi}_s^e = \left(\xi_0, \xi_0^e, \xi_s^0, \xi_s^e\right)$ are e.g. kinetic parameters or initial conditions. Here, ξ_0 are the parameters that are the same for all conditions and subpopulations, ξ_0^e and ξ_s^0 the parameters that are different across experiments or subpopulations, respectively, and ξ_s^e the parameters that differ between experiments and subpopulation. We added the index e to the subpopulation parameters $\boldsymbol{\psi}_s^e$ and the mixture weights w_s^e to allow the parameter to differ between experiments, since we use ODE-MMs to detect differences between experimental conditions. The system is stimulated with an external, possibly experiment specific stimulus denoted by u^e. For an illustration of the method see Figure 3.1.

3.1.2 Problem Statement

Despite the successes achieved using ODE-MM, there are several open issues. So far, only the means of the subpopulations are described by ODEs and the variances are treated as additional parameters. If many time points and experimental conditions are observed, the unknown variances increase the dimension of the parameter space significantly. As the predictive power of a model generally decreases with its complexity, it is desirable to reduce the number of its parameters. Moreover, RREs provide only a description of the averaged behavior of a subpopulation. They are neither able to describe intrinsic noise, arising due to stochasticity of births and deaths of single molecules, nor extrinsic noise, emerging from stochastic variability of parameters. Furthermore, it is not possible to exploit correlation structures among multivariate measurements, as they only can be analyzed independently. Even if subpopulations can be identified, correlation structures among the measurements may not be detected analyzing one measurement at a time. In particular, this chapter addresses the following problems:

Problem 1 High dimension of the parameter space.

Problem 2 No mechanistic description of intrinsic variability of subpopulations.

Problem 3 No accounting for extrinsic noise in a subpopulation.

Problem 4 No consideration and detection of correlations between multivariate measurements.

Problem 5 Numerical instability arising due to mixture modeling.

In the following, we will address these problems by extending ODE-MM. Furthermore, as there are merely two assessments of ODE-MM, will provide additional evaluations on artificial as well as real experimental data.

3.2 Assessment of ODE-MMs Using Novel Data for NGF-Induced Erk Signaling

Understanding intracellular signaling mechanisms that regulate pain sensitization is of great importance for pain research. Therefore, the underlying processes of NGF-induced Erk phosphorylation are studied. ODE-MMs have been used to investigate this pathway

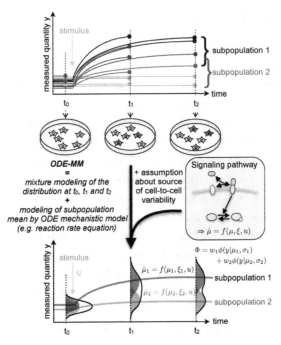

Figure 3.1: Illustration of ODE constrained mixture modeling. The combination of mixture modeling of experimental data and pathway information allows us improved aquisition of subpopulation structures and mechanistics. This figure has been adopted from (Hasenauer *et al.*, 2014b).

in primary sensory neurons (Hasenauer *et al.*, 2014b). These cells are used to study pain sensitization and provide a suitable application for ODE-MM due to their high heterogeneity.

In this section we evaluate the usage of ODE-MMs to study not only differences among individual subpopulations, but also differences between experimental conditions. We investigate the alteration of subpopulation response by cellular environment. Therefore, we describe a simple pathway model of NGF-induced Erk activation introduced by Hasenauer *et al.* (2014b), which builds the basis for further analysis. We analyze experimental data of NGF-induced Erk signaling that has been generated under several experimental conditions by Katharina Möller and Tim Hucho[1]. Here, we present the analysis of two of these conditions by applying ODE-MMs with RREs with the aim to detect the source of difference between the conditions.

3.2.1 Pathway Model for NGF-Induced Erk Phosphorylation

The pathway model of NGF-induced Erk phosporylation proposed by Hasenauer *et al.* (2014b) states that binding of NGF and the receptor TrkA results in a complex TrkA:NGF, which induces phosporylation of Erk. This process can be described with the reactions

$$R_1 : \quad \text{TrkA} + \text{NGF} \to \text{TrkA:NGF}, \qquad \text{rate} = k_1[\text{TrkA}][\text{NGF}],$$
$$R_2 : \quad \text{TrkA} \to \text{TrkA} + \text{NGF}, \qquad \text{rate} = k_2[\text{TrkA:NGF}],$$
$$R_3 : \quad \text{Erk} \to \text{pErk}, \qquad \text{rate} = k_3[\text{TrkA:NGF}][\text{Erk}],$$
$$R_4 : \quad \text{Erk} \to \text{pErk}, \qquad \text{rate} = k_4[\text{Erk}],$$
$$R_5 : \quad \text{pErk} \to \text{Erk}, \qquad \text{rate} = k_5[\text{pErk}].$$

Assuming conservation of mass yields

$$[\text{TrkA}] + [\text{TrkA:NGF}] = [\text{TrkA}]_0,$$
$$[\text{NGF}] + [\text{TrkA:NGF}] = [\text{NGF}]_0,$$
$$[\text{Erk}] + [\text{pErk}] = [\text{Erk}]_0.$$

To eliminate structurally non-identifiable parameters, the model can be reparametrized. The final RREs are (see (Hasenauer *et al.*, 2014b) for further details)

$$\frac{dx_1}{dt} = k_1[\text{NGF}]_0(k_3[\text{TrkA}]_0 - x_1) - k_2 x_1,$$
$$\frac{dx_2}{dt} = (x_1 + k_4)(s[\text{Erk}]_0 - x_2) - k_5 x_2,$$
$$y = x_2,$$

with $x_1 = k_3[\text{TrkA:NGF}]$ and $x_2 = s[\text{pErk}]$. The measurand $y = s[\text{pErk}]$ can only be measured up to some scaling constant s. This pathway model has been studied using

---

[1]Division of Experimental Anesthesiology and Pain Research at the Department of Anesthesiology and Intensive Care Medicine at the University Hospital Cologne

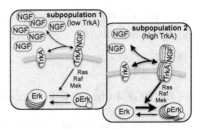

Figure 3.2: Model for NGF-induced Erk phosphorylation. Low and high responsiveness of the subpopulations to NGF stimulation can be explained by different levels of the NGF receptor TrkA. The signaling intermediates Ras, Raf and Mek are not modeled in this simple pathway model. Thicker arrows, corresponding to the influence of TrkA:NGF on the Erk phosphorylation, and the higher abundance of TrkA in subpopulation 2, visualize the difference between the subpopulations. This figure has been adopted from (Hasenauer et al., 2014b).

several ODE-MMs with RREs based on single-cell data of NGF-stimulated primary sensory neurons. This revealed that the cell population consists of two subpopulations that differ in TrkA levels and therefore show a different response to NGF stimulation (see Figure 3.2).

3.2.2 Experimental Data and Problem Statement

A goal of pain research is to fully understand the mechanism of pain sensitization. Therefore, it is studied how different conditions influence the pathways mediating pain-sensitivity. The data analyzed in this section is generated under two different experimental conditions and each experiment has been repeated three times. Cells are stimulated with NGF and the concentration of pErk is measured after 1, 5, 30, 60 and 120 minutes. The data is visualized in Figure 3.3, with histograms for every time point (Figure 3.3A), the average concentrations of pErk (Figure 3.3B) and the mean number of cells per time point (Figure 3.3C). Two main differences between the conditions can be observed:

- The mean concentration of pErk is lower for condition 1.

- The number of cells is higher under condition 1.

These observations give rise to the question, where the differences in average pErk concentrations comes from: *Is the relative size of the responsive subpopulation higher under condition 2 or does it have a higher response to NGF stimulation?*

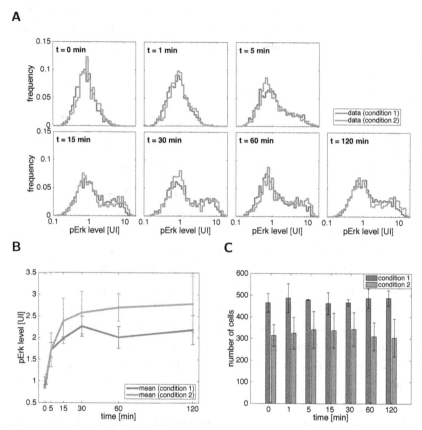

Figure 3.3: Snapshot data of NGF-induced Erk signaling under two different experimental conditions. (**A**) Histograms of experimental data for 7 time points. (**B**) Mean and standard deviation of pErk levels for three biological replicates. Levels of pErk are given in arbitrary units of intensity [UI]. The average amount of pErk is higher under condition 2. (**C**) The mean over the number of cells per time point for every replicate shows that there are more cells under condition 1.

3.2.3 Hypothesis Testing

We perform hypothesis testing to answer the question of the source of difference between the experimental conditions. Hasenauer *et al.* (2014b) showed that at least two subpopulations are present that differ in TrkA levels. Therefore, we assume a subpopulation structure, which arises due to differences in TrkA levels for the following hypotheses (see Figure 3.4):

H1 No difference between the conditions.

H2 Higher relative size of the high responsive subpopulation under condition 2.

H3 Higher response to NGF stimulation of the high responsive cells under condition 2.

H4 Higher response to NGF stimulation of both subpopulations under condition 2.

In accordance to the optimal model selected in the studies of Hasenauer *et al.* (2014b), we assume a log-normal distribution parameterized by the median of the subpopulations. H2 explains the difference by assuming that less cells of the low responsive subpopulation exist in condition 2 and therefore the average concentration of pErk is higher. Under H3 and H4, the weighting of the subpopulations stays the same, but the response to NGF stimulation is changed in condition 2. While H3 only allows a higher response for the high responsive subpopulations, the responsiveness for all cells is higher under H4. H2 considers different weightings for the experimental condition, while H3 and H4 include different responses to stimulation with NGF. The higher response is modeled by multiplying parameter $k_3[\text{TrkA}]_0$ by a parameter κ, which describes the stimulus-dependent response.

To obtain estimates of the parameters we perform multi-start local optimization with 100 multi-starts. If the optimizer finds the same (possibly local) optimum less than 5 times, we increase the number of multi-starts and repeat the optimization. We restrict the kinetic parameters to the interval $[10^{-10}, 10^{10}]$, the variances to $[10^{-1}, 10]$, the weights to $[0, 1]$ and the additional parameter κ, which is used in H3 and H4 to $[10^{-10}, 10^{10}]$. Model selection using AIC and BIC selects H3 and H4 for the pooled data, as shown in Table 3.1. The fitted data of the optimal model is depicted in Figure 3.5. Repeating the procedure for the single replicates shows that the significance is not as high as for the pooled data, but nevertheless, H4 is not rejected for any replicate (see last columns of Table 3.1).

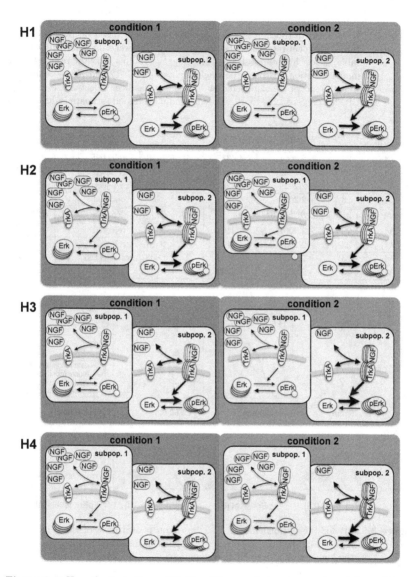

Figure 3.4: Hypothesis testing: **H1** No difference between experimental conditions. **H2** Higher relative size of responsive subpopulation for condition 2. **H3** Higher response to NGF stimulation of the high responsive cells for condition 2. **H4** Higher response to NGF stimulation of both subpopulations for condition 2. Differences are visualized by thickness of arrows and abundance of species.

Table 3.1: Hypothesis testing for two experimental condition based on pooled data of three biological replicates. Both criteria, AIC (lower table) and BIC (upper table), select hypotheses H3 and H4. The last colums show the results for the model selection based on the single replicates $\mathcal{R}1$, $\mathcal{R}2$ and $\mathcal{R}3$. ✓ indicates that the model is not rejected and ✗ that it has been rejected using AIC or BIC for Δ_{BIC} or $\Delta_{\mathrm{AIC}} > 10$. The maximum likelihood estimate is denoted by $\boldsymbol{\theta}^{\mathrm{ML}}$.

hypothesis	n_θ	$\log L(\boldsymbol{\theta}^{\mathrm{ML}})(10^4)$	BIC (10^4)	rank	Δ_{BIC}	decision	$\mathcal{R}1$	$\mathcal{R}2$	$\mathcal{R}3$
$\mathcal{M}_{\mathrm{H1}}$	22	-2.2061	4.4337	4	> 10	rejected	✓	✓	✗
$\mathcal{M}_{\mathrm{H2}}$	23	-2.2052	4.4329	3	> 10	rejected	✗	✗	✓
$\mathcal{M}_{\mathrm{H3}}$	23	-2.2029	4.4281	2	3.808	not rejected	✓	✓	✗
$\mathcal{M}_{\mathrm{H4}}$	23	-2.2027	4.4277	1	0	optimal	✓	✓	✓
hypothesis	n_θ	$\log L(\boldsymbol{\theta}^{\mathrm{ML}})(10^4)$	AIC (10^4)	rank	Δ_{AIC}	decision	$\mathcal{R}1$	$\mathcal{R}2$	$\mathcal{R}3$
$\mathcal{M}_{\mathrm{H1}}$	22	-2.2061	4.4167	4	> 10	rejected	✗	✓	✓
$\mathcal{M}_{\mathrm{H2}}$	23	-2.2052	4.4151	3	> 10	rejected	✗	✓	✓
$\mathcal{M}_{\mathrm{H3}}$	23	-2.2029	4.4103	2	3.808	not rejected	✓	✓	✓
$\mathcal{M}_{\mathrm{H4}}$	23	-2.2027	4.4099	1	0	optimal	✓	✓	✓

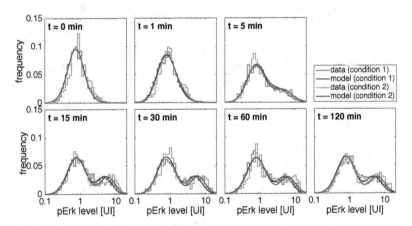

Figure 3.5: Fit for the optimal model M_{H4} based on pooled data of three biological replicates. The high responsive subpopulation is shifted to the right, due to the higher response to NGF stimulation.

In summary, we assessed how ODE-MMs with RREs can be applied to analyze data obtained under different experimental conditions simultaneously at the example of NGF-induced Erk signaling. We formulated several hypotheses about the difference between conditions that yields a difference in mean levels of pErk. To test the hypotheses we performed parameter estimation and model selection using a simple pathway model of NGF-induced phosphorylation of Erk. Based on pooled data consisting of three replicates, both BIC and AIC select models that consider a changed intracellular signaling under the second condition. These models incorporate a higher response to NGF stimulation under condition 2. A model that assumes a different weighting under the conditions, which was motivated by the fact that the number of cells differs significantly under the two conditions, has been rejected based on the pooled data. Repeating model selection for the single replicates, the model, which assumes a higher phosphorylation of Erk in both subpopulations under condition 2, can not be rejected for any replicate neither by AIC nor by BIC. In the future, we will further analyze dose-response data to validate the results of the model selection.

3.3 Modeling Variability within a Subpopulation

Considering only the averaged behavior of a cell population might not represent the behavior of cells at the tail of the cell distribution (Altschuler & Wu, 2010). This gets even worse for cell populations that consist of two or more subpopulations and therefore show a bimodal or multimodal distribution of the cells. In addition, it has been shown that variability of measured properties carries information about the underlying biological system (Munsky et al., 2009). These facts motivate the usage of a mechanistic description of the variability of a system.

The previously introduced method of ODE-MMs with RREs accounts for differences between subpopulations (Hasenauer et al., 2014b). Nevertheless, modeling subpopulation dynamics by RREs gives no mechanistic description of the cell-to-cell variability within a subpopulation. A possibility to exploit higher order moments of the subpopulations is to describe the dynamics of a subpopulation by moment equations (MEs) (Engblom, 2006). In this section we use ODE-MMs with MEs to solve the Problems 1-3 that have been addressed in Section 3.1.2. First, we propose in Section 3.3.1 a likelihood function for ODE-MMs with MEs to study univariate measurements $y \in \mathbb{R}$. Additionally, we describe how the MEs can be linked to a normal and log-normal mixture distribution.

In Section 3.3.2, we validate the method for different scenarios of a conversion process. Besides that, we compare the results of the method with those obtained using RREs for the description of the mechanisms of the system.

3.3.1 Likelihood Function

The likelihood function of mixture modeling that is constrained by MEs is given by

$$L(\boldsymbol{\theta}) = \prod_{e,k,j} \sum_{s=1}^{n_s} w_s^e \, p\left(\bar{y}_j^{e,k} | \boldsymbol{\varphi}_s^e(t_k)\right) \tag{3.2}$$

$$\text{with} \quad \dot{\mathbf{x}}_s^e = f(\mathbf{x}_s^e, \boldsymbol{\psi}_s^e, u^e), \quad \mathbf{x}_s^e(0) = \mathbf{x}_0(\boldsymbol{\psi}_s^e, u^e),$$

$$\boldsymbol{\varphi}_s^e = h(\mathbf{x}_s^e, \boldsymbol{\psi}_s^e, u^e).$$

In the following we neglect the indices k for the time point, and j for the single-cells. The likelihood describes the probability of observing the measurement $\bar{y}^e \in \mathbb{R}$ as weighted sum of mixture probabilities $p(\bar{y}|\boldsymbol{\varphi}_s^e)$ with parameters $\boldsymbol{\varphi}_s^e$ for subpopulation s in experiment e. Each of the n_s subpopulations has a weight denoted by w_s^e corresponding to its size. The parameters $\boldsymbol{\psi}_s^e = \left(\xi_0, \xi_s^e, \xi_s^0, \xi_s^e\right)$ are e.g. kinetic parameters or initial conditions that are partitioned into experiment specific and identical parameters between conditions as in (3.1). The cells are stimulated with an experiment specific external stimulus denoted by u^e. The time evolution of the moments $\mathbf{x}_s^e$ of the system are described by a function f. The moments can be linked to the mixture parameters with function h. We also neglect the indices e and s in the following. Since a mechanistic description of the variability is provided by the MEs, measurement noise e.g. normal additive measurement noise or log-normal multiplicative measurement noise

$$\bar{y} = y + \epsilon, \ \epsilon \sim \mathcal{N}(0, \sigma_\epsilon^2), \tag{3.3}$$

$$\bar{y} = y\epsilon, \quad \epsilon \sim \log \mathcal{N}(0, \sigma_\epsilon^2), \tag{3.4}$$

can be considererd separately.

In this thesis, we consider second order moments. The state vector of the moments comprises mean and variances of the L species:

$$\mathbf{x} = \begin{pmatrix} \mathbf{m} \\ \mathbf{C} \end{pmatrix}, \quad \mathbf{m} = (m_1, \dots, m_L), \quad (\mathbf{C})_{ij} = C_{ij}, \; i,j = 1, \dots, L.$$

If we assume to have at most quadratic propensities $a_r(\mathbf{m})$ for the M reactions and if we neglect higher order moments, the MEs are (Engblom, 2006)

$$\frac{dm_i}{dt} = \sum_{r=1}^{M} \nu_{ir} \left(a_r(\mathbf{m}) + \frac{1}{2} \sum_{l_1, l_2} \frac{\partial^2 a_r(\mathbf{m})}{\partial x_{l_1} \partial x_{l_2}} C_{l_1 l_2} \right),$$

$$\frac{dC_{ij}}{dt} = \sum_{r=1}^{M} \left(\nu_{ir} \sum_l \frac{\partial a_r(\mathbf{m})}{\partial x_l} C_{jl} + \nu_{jr} \sum_l \frac{\partial a_r(\mathbf{m})}{\partial x_l} C_{il} + \right.$$
$$\left. \nu_{ir} \nu_{jr} \left(a_r(\mathbf{m}) + \frac{1}{2} \sum_{l_1, l_2} \frac{\partial^2 a_r(\mathbf{m})}{\partial x_{l_1} \partial x_{l_2}} C_{l_1 l_2} \right) \right),$$

with ν_{ij} being the entries of the stochiometric matrix. These moments can be linked to the measurand without measurement noise y to obtain its mean m_y and variance C_y.

Example. If the output is given by $y = bx_l$, i.e., y is proportional to the amount of x_l, we obtain the mean m_l and the variance C_{ll} from the corresponding entries of the state vector of the moments. Thus, we can calculate the mean of the output $m_y = bm_l$ and its variance $C_y = b^2 C_{ll}$.

By describing subpopulation dynamics with MEs, extrinsic noise of the cells in a subpopulation that arises due to differences e.g. in kinetic parameters, can be incorporated. The variable parameters can be defined as states and corresponding moment equations can be derived and simulated. We use normal and log-normal mixture distributions defined by its mixture parameters $\varphi_s^e = (\mu_s^e, \sigma_s^e)$. Using second order moments, we can link both parameters to the ME, which we will explain in the following. However, this yields a reduced number of unknown parameters $\boldsymbol{\theta} = \{\{w_s^e, \psi_s^e\}_s, \sigma_\epsilon^e\}_e$, since no additional parameters for the variances needs to be introduced and estimated as in (3.1).

Mixture of Normal Distributions

Both moments of the measurand, m_y and C_y and their corresponding sensitivities $\frac{dm_y}{d\theta}, \frac{dC_y}{d\theta}$ can directly be linked to mixture parameters of a normal distribution by

$$\mu = m_y, \quad \sigma^2 = C_y + \sigma_\epsilon^2, \quad \frac{d\mu}{d\theta} = \frac{dm_y}{d\theta} \quad \text{and} \quad \frac{d\sigma}{d\theta} = \frac{1}{2\sigma} \left(\frac{dC_y}{d\theta} + \frac{d\sigma_\epsilon^2}{d\theta} \right).$$

Here, σ_ϵ^2 is the variance of additive normally distributed measurement noise (3.3).

Mixture of Log-Normal Distributions

Another mixture distribution used in this thesis is the log-normal distribution. If the mean of a log-normal distribution is linked to the mean of the moment equations, we need to following calculations:

$$\mu = \log(m_y) - \frac{1}{2}\sigma^2,$$

$$\sigma^2 = \log\left(1 + \frac{C_y}{m_y^2}\right) + \sigma_\epsilon^2,$$

$$\frac{d\mu}{d\theta} = \frac{1}{m_y}\frac{dm_y}{d\theta} - \frac{1}{2}\frac{d\log\left(\frac{C_y}{m_y^2}+1\right)}{d\theta} - \frac{1}{2}\frac{d\sigma_\epsilon^2}{d\theta},$$

$$\frac{d\sigma}{d\theta} = \frac{1}{2\sigma}\left(\frac{d\log(\frac{C_y}{m_y^2}+1)}{d\theta} + \frac{d\sigma_\epsilon^2}{d\theta}\right),$$

$$\text{with} \quad \frac{d\log\left(\frac{C_y}{m_y^2}+1\right)}{d\theta} = \frac{1}{\frac{C_y}{m_y^2}+1}\frac{m_y^2\frac{dC_y}{d\theta} - 2C_y m_y\frac{dm_y}{d\theta}}{m_y^4}.$$

Here, σ_ϵ^2 is the mixture parameter of multiplicative log-normally distributed measurement noise (3.4). The median of the distribution can be described in a similar way by the means of the MEs.

3.3.2 Simulation Example: Conversion Reaction

To assess the method of ODE-MMs with MEs we study a conversion reaction between some biochemical species A and B, a frequently occurring process in biology. A schematic of the process is shown in Figure 3.6. The reaction system describing this process is

$$\begin{aligned}
R_1: \quad & A \to B, \quad \text{rate} = k_1 u [A], \\
R_2: \quad & A \to B, \quad \text{rate} = k_2 [A], \\
R_3: \quad & B \to A, \quad \text{rate} = k_3 [B].
\end{aligned}$$

For the conversion of A to B we denote a time dependent stimulus by u and distinguish two different reactions. First, a stimulus dependent reaction R_1 occurring with rate $k_1 u [A]$, where $[A]$ denotes the concentration of species A, and second, a basal, stimulus independent reaction R_2. In reaction R_3 species B is converted to A with kinetic parameter k_3. Due to conservation of mass, the sum of concentrations $[A] + [B]$ remains constant.

Artificial Data

We generate artificial data for an external stimulus $u(t) = 0$ for $t \leq 0$ and $u(t) = 1$ for $t > 0$, i.e., the system is in steady state without stimulus at initial time. We generate trajectories of 1000 cells in a volume $\Omega = 1000$, which have a total number of molecules $N_0 = 1000$ using the SSA. A certain fraction of the cells shows higher response to stimulus u. This difference is modeled with the subpopulation specific parameter k_1. For every time series we measure B at $t = 0, 0.1, 0.2, 0.3, 0.5$ and 1 minutes. As the use of MEs also provides information about variability in parameters we consider three scenarios:

Scenario 1 Kinetic parameters are fixed across individual cells of a subpopulation.

Scenario 2 Kinetic parameters vary little between individual cells of a subpopulation.

Scenario 3 Kinetic parameters vary strongly between individual cells of a subpopulation.

The scenarios are depicted in Figure 3.7. For every scenario we simulate two subpopulations that respond differently to stimulus u, i.e., one subpopulation has a stimulus-

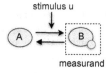

Figure 3.6: Schematic representation of a conversion process between species A and B, for which B can be measured. This figure has been adopted from (Hasenauer *et al.*, 2014b).

dependent conversion with parameter $k_{1,s1} = 0.75$, while the other shows a lower response with parameter $k_{1,s2} = 0.1$. For Scenarios 2 and 3 we assume that the parameters are log-normally distributed with mean μ_{k_i} and variance $\sigma^2_{k_i}$. The variance for Scenario 2 ($\sigma^2_{k_i} = 0.0016$) is smaller than for Scenario 3 ($\sigma^2_{k_i} = 0.005$). We want to emphasize that by μ_{k_i} and $\sigma^2_{k_i}$, we denote the means and variances of the parameters instead of the corresponding parameters of the log-normal distribution.

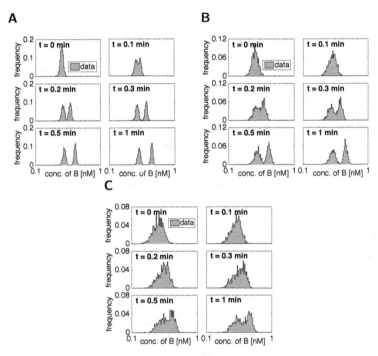

Figure 3.7: Artificial data for different scenarios of a conversion process of A and B with two subpopulations that differ in the response to a stimulus: Scenarios with (**A**) no parameter variability, (**B**) low parameter variability and (**C**) high parameter variability between individual cells.

Moment Equations for The Conversion Process without Variability in Parameters

The MEs for the conversion process are generated and simulated using the toolbox CER-ENA developed by Kazeroonian et al. (2016). Given the system size $\Omega = 1000$, we obtain the following equations for the mean and variance of the concentration of B

$$\frac{dm_B}{dt} = -k_3 m_B - (k_1 + k_2)(m_B - 1),$$

$$\frac{dC_{B,B}}{dt} = \frac{k_3 m_B - (k_1 + k_2)(m_B - 1)}{\Omega} - 2C_{B,B}(k_1 + k_2 + k_3).$$

Based on these equations and the initial steady state assumption, we derive the initial conditions for the MEs

$$m_B(0) = \frac{k_2}{k_2 + k_3} \quad \text{and} \quad C_{B,B}(0) = \frac{k_2 k_3}{\Omega(k_2 + k_3)^2}.$$

Moment Equations for The Conversion Process with Variability in Parameters

The MEs for the conversion process with accounting for additional variability in the parameters are presented in the following. Given the means μ_{k_i} and standard deviations σ_{k_i} of the parameters for $i = 1, 2, 3$, we obtain

$$\frac{dm_B}{dt} = (m_B - 1)(m_{k_3} - m_{k_1}) + m_B m_{k_2} - C_{B,k_1} + C_{B,k_2} + C_{B,k_3},$$

$$\frac{dm_{k_i}}{dt} = 0 \text{ for } i = 1, 2, 3,$$

$$\frac{dC_{B,B}}{dt} = \frac{(m_B - 1)(m_{k_3} - m_{k_1}) + m_B m_{k_2} - C_{B,k_1} + C_{B,k_2} + C_{B,k_3}}{\Omega} +$$
$$2(C_{B,k_3} - C_{B,k_1})(m_B - 1) + 2C_{B,k_2} m_B + 2C_{B,B}(m_{k_3} + m_{k_2} - m_{k_1}),$$

$$\frac{dC_{B,k_1}}{dt} = C_{B,k_1}(m_{k_2} + m_{k_3} - m_{k_1}) + C_{k_1,k_2} m_B - C_{k_1,k_1}(m_B - 1) + C_{k_1,k_3} m_B - 1,$$

$$\frac{dC_{B,k_2}}{dt} = C_{B,k_2}(m_{k_2} + m_{k_3} - m_{k_1}) + C_{k_2,k_2} m_B - C_{k_1,k_2}(m_B - 1) + C_{k_2,k_3} m_B - 1,$$

$$\frac{dC_{B,k_3}}{dt} = C_{B,k_3}(m_{k_2} + m_{k_3} - m_{k_1}) + C_{k_2,k_3} m_B - C_{k_1,k_3}(m_B - 1) + C_{k_3,k_3} m_B - 1,$$

$$\frac{dC_{k_i,k_j}}{dt} = 0 \text{ for } i, j = 1, 2, 3.$$

By exploiting the steady state assumption, we obtain the initial conditions

$$m_B(0) = \Omega \frac{\mu_{k_2}^2 + \mu_{k_2}\mu_{k_3} - \sigma_{k_2}^2}{\mu_{k_2}^2 + \mu_{k_2}\mu_{k_3} + \mu_{k_3}^2 - \sigma_{k_2}^2 - \sigma_{k_3}^2},$$

$$m_{k_i}(0) = \mu_{k_i},$$

$$C_{B,k_1}(0) = 0,$$

$$C_{B,k_2}(0) = \frac{\Omega \sigma_{k_2}^2 (\mu_{k_3}^2 + \mu_{k_3}\mu_{k_2} - \sigma_{k_3}^2)}{(\mu_{k_2} + \mu_{k_3})(\mu_{k_2}^2 + 2\mu_{k_2}\mu_{k_3} + \mu_{k_3}^2 - \sigma_{k_2}^2 - \sigma_{k_3}^2)},$$

$$C_{B,k_3}(0) = \frac{-\Omega \sigma_{k_3}^2 (\mu_{k_2}^2 + \mu_{k_3}\mu_{k_2} - \sigma_{k_2}^2)}{(\mu_{k_2} + \mu_{k_3})(\mu_{k_2}^2 + 2\mu_{k_2}\mu_{k_3} + \mu_{k_3}^2 - \sigma_{k_2}^2 - \sigma_{k_3}^2)},$$

$$C_{B,B}(0) = \frac{(\Omega - 1)C_{B,k_2}(0) + C_{B,k_3}(0) + \Omega\mu_{k_2}}{2(\mu_{k_2} + \mu_{k_3})} + $$
$$\frac{m_B(0)(\mu_{k_3} - \mu_{k_2}) - 2C_{B,k_2}(0) - 2C_{B,k_3}(0)}{2(\mu_{k_2} + \mu_{k_3})},$$

$$C_{k_i,k_i}(0) = \sigma_{k_i}^2.$$

Hypothesis Testing for Scenario 1

To assess the extension and compare ODE-MM with RREs and with MEs, we perform hypothesis testing for both methods based on the generated data of Scenario 1 (see Figure 3.7A):

H1 No subpopulations.

H2 Two subpopulations differing in k_1.

H3 Two subpopulations differing in k_2.

H4 Two subpopulations differing in k_3.

The kinetic parameters are restricted to the interval $[10^{-6}, 10^4]$ and the weighting of the subpopulations needed for H2-4 to the interval $[0, 1]$. As the variances need to be estimated when using RREs, we restrict them to $[10^{-2.5}, 10^{2.5}]$. Moreover, we use three different distributions assumptions for every hypothesis, a normal distribution, for which the mean is described by the ODEs, and a log-normal distribution, for which either the mean or the median is parameterized by the ODE model. This yields 12 models that are tested with multi-start local optimization and model selection using AIC and BIC.

We perform parameter estimation with a toolbox that is internally used by the Data-driven Computational Modeling group of the Institute of Computational Biology at the Helmholtz Zentrum München. The results are shown in Table 3.2 for ODE-MMs with RREs and in Table 3.3 for ODE-MMS with MEs. Both select the same optimal model, which detects the true differences between the subpopulations. The fit of the optimal model $\mathcal{M}_{H2,1}$ is shown in Figure 3.8A. Furthermore, we computed the profile likelihoods of the kinetic parameters and weights of the optimal models, which are shown in Figure 3.8B. All parameters are identifiable and the profiles are almost indistinguishable. The number of parameters can be reduced by at least a factor of two using MEs with $n_{\theta} = 5$ for MEs and $n_{\theta} = 17$ using RREs. Moreover, model selection between ODE-MM with RRE and ME yields $\text{BIC}_{\min}^{\text{RRE}} - \text{BIC}_{\min}^{\text{ME}} > 10$, i.e., the model using MEs is selected in favor of the RRE model.

Measurements at Less Time Points in Scenario 1

Additionally, we compare the performance of ODE-MMs with ME and RRE for the case that less measurements are available. Thus, we repeat the parameter estimation for the optimal model based on only three time points $t = 0, 0.1$ and 0.5 minutes. As the MEs also can extract information from the variance of the subpopulation, the confidence intervals are narrower for the case of using MEs instead of RREs (see Figure 3.8C). We expect that for some other systems the uncertainties of the parameters for ODE-MMs with MEs is much lower than using RREs. Munsky et al. (2009) already showed for some processes that measurements at less time points are needed to obtain identifiable parameters if second order moments are measured besides the means.

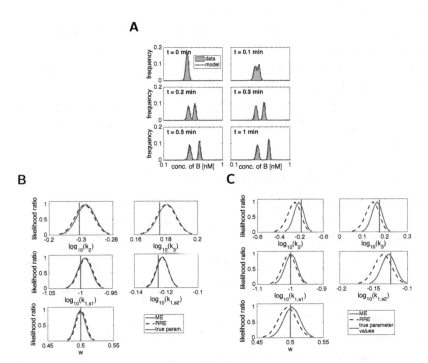

Figure 3.8: Results for Scenario 1. (**A**) Fitted data of the optimal model $\mathcal{M}_{H2,1}$ using ODE-MMs with MEs. (**B**, **C**) Comparison of profile likelihoods for ODE-MMs with RREs (red line) and MEs (dotted dark red line) for different numbers of measurements. (**B**) For the case of 7 time points, almost no difference can be detected between the profiles. (**C**) If measurements of less time points ($t = 0, 0.1, 0.5$ min) are available, ODE-MMs with ME yields higher confidence in the estimates.

Table 3.2: Results of parameter estimation and model selection for Scenario 1 using ODE-MMs with RREs.

	n_s	distribution	ODE const.	diff.	n_θ	$\log L(\boldsymbol{\theta}^{\mathrm{ML}})(10^4)$	BIC (10^4)	rank	Δ_{BIC}	decision
$\mathcal{M}_{\mathrm{H}1,1}$	1	normal	mean	-	9	1.0821	-2.1563	12	> 10	rejected
$\mathcal{M}_{\mathrm{H}1,2}$	1	log-normal	mean	-	9	1.0837	-2.1597	10	> 10	rejected
$\mathcal{M}_{\mathrm{H}1,3}$	1	log-normal	median	-	9	1.0837	-2.1596	11	> 10	rejected
$\mathcal{M}_{\mathrm{H}2,1}$	2	normal	mean	k_1	17	1.3562	-2.6975	1	0	optimal
$\mathcal{M}_{\mathrm{H}2,2}$	2	log-normal	mean	k_1	17	1.3556	-2.6963	2	> 10	rejected
$\mathcal{M}_{\mathrm{H}2,3}$	2	log-normal	median	k_1	17	1.3556	-2.6963	3	> 10	rejected
$\mathcal{M}_{\mathrm{H}3,1}$	2	normal	mean	k_2	17	1.1233	-2.2318	8	> 10	rejected
$\mathcal{M}_{\mathrm{H}3,2}$	2	log-normal	mean	k_2	17	1.1296	-2.2445	7	> 10	rejected
$\mathcal{M}_{\mathrm{H}3,3}$	2	log-normal	median	k_2	17	1.1208	-2.2267	9	> 10	rejected
$\mathcal{M}_{\mathrm{H}4,1}$	2	normal	mean	k_3	17	1.1968	-2.3787	4	> 10	rejected
$\mathcal{M}_{\mathrm{H}4,2}$	2	log-normal	mean	k_3	17	1.1855	-2.3563	6	> 10	rejected
$\mathcal{M}_{\mathrm{H}4,3}$	2	log-normal	median	k_3	17	1.1900	-2.3652	5	> 10	rejected

Table 3.3: Results of parameter estimation and model selection for Scenario 1 using ODE-MMs with MEs.

	n_s	distribution	ODE const.	diff.	n_θ	$\log L(\boldsymbol{\theta}^{\mathrm{ML}})(10^4)$	BIC (10^4)	rank	Δ_{BIC}	decision
$\mathcal{M}_{\mathrm{H}1,1}$	1	normal	mean	-	3	-1.6347	3.2721	12	> 10	rejected
$\mathcal{M}_{\mathrm{H}1,2}$	1	log-normal	mean	-	3	-1.5787	3.1600	11	> 10	rejected
$\mathcal{M}_{\mathrm{H}1,3}$	1	log-normal	median	-	3	-1.5736	3.1498	10	> 10	rejected
$\mathcal{M}_{\mathrm{H}2,1}$	2	normal	mean	k_1	5	1.3556	-2.7069	1	0	optimal
$\mathcal{M}_{\mathrm{H}2,2}$	2	log-normal	mean	k_1	5	1.3550	-2.7057	3	> 10	rejected
$\mathcal{M}_{\mathrm{H}2,3}$	2	log-normal	median	k_1	5	1.3551	-2.7058	2	> 10	rejected
$\mathcal{M}_{\mathrm{H}3,1}$	2	normal	mean	k_2	5	0.7804	-1.5564	9	> 10	rejected
$\mathcal{M}_{\mathrm{H}3,2}$	2	log-normal	mean	k_2	5	0.7920	-1.5796	8	> 10	rejected
$\mathcal{M}_{\mathrm{H}3,3}$	2	log-normal	median	k_2	5	0.7928	-1.5813	7	> 10	rejected
$\mathcal{M}_{\mathrm{H}4,1}$	2	normal	mean	k_3	5	1.0548	-2.1053	4	> 10	rejected
$\mathcal{M}_{\mathrm{H}4,2}$	2	log-normal	mean	k_3	5	1.0531	-2.1019	6	> 10	rejected
$\mathcal{M}_{\mathrm{H}4,3}$	2	log-normal	median	k_3	5	1.0541	-2.1038	5	> 10	rejected

Hypothesis Testing for Scenario 2 and 3

For Scenario 2 and 3 we assess the ability of ODE-MMs with MEs to detect extrinsic noise in a subpopulation. We perform hypothesis testing based on data that has been generated with variable parameters (see Figures 3.7B and C) for the following hypotheses

H1 Two subpopulations that differ in k_1 and fixed parameters between individual cells of a subpopulation.

H2 Two subpopulations that differ in k_1 and variability of parameters between individual cells of a subpopulation.

In this case the ground truth is H2, which assumes an additional cell-to-cell variability. We model the data using ODE-MMs with MEs and a normal mixture distribution that is constrained by its mean. For parameter estimation for H1 we use the same restrictions in parameters as before. For the models of H2 the means of the kinetic parameters are restricted to the interval $[10^{-6}, 10^4]$ and the coefficient of variation of the parameters $cv_{k_i} = \sigma_{k_i}/\mu_{k_i}$ to $[10^{-6}, 10^0]$.

The results are shown in Table 3.4 for low parameter variabilty (Scenario 2) and in Table 3.5 for high parameter variability (Scenario 3). Our method is able to detect the true model with high significance for both cases. Figures 3.9A and B show the fits of model $\mathcal{M}_{\mathbf{H1}}$, which allows no additional variability in parameters, and $\mathcal{M}_{\mathbf{H2}}$, which accounts for cell-to-cell variability in kinetic parameters. The fits of M_{H1} M_{H2} shows the importance of accounting for extrinsic noise. The profiles corresponding to the optimal model are shown in Figures 3.9C and D. The subpopulation weight and the means of the kinetic parameters are identifiable. For the case of low parameter variability also the coefficient of variation of k_3, cv_{k_3}, can be estimated with high confidence. As the MEs give no exact representation of the system, it can happen that the true parameters lie outside the confidence intervals. This is the case for cv_{k_2} and $cv_{k_{1,s1}}$ in Scenario 2 and cv_{k_2}, cv_{k_3} and $cv_{k_{1,s2}}$ for Scenario 3. As expected, the results are better for lower variability in parameters.

In summary, we combined ODE constrained mixture modeling with MEs for the description of the mechanisms of a biological process. As variances are linked to the MEs the

Table 3.4: Model selection for data of Scenario 2 that includes low parameter variability. Both criteria, AIC and BIC, select the true model $\mathcal{M}_{H2}$, which allows parameters to vary between individual cells.

	n_θ	$\log L(\theta^{\mathrm{ML}})(10^4)$	AIC (10^4)	rank	Δ_{AIC}	decision
$\mathcal{M}_{H1}$	5	0.6504	-1.2998	2	> 10	rejected
$\mathcal{M}_{H2}$	9	1.0280	-2.0542	1	0	optimal
	n_θ	$\log L(\theta^{\mathrm{ML}})(10^4)$	BIC (10^4)	rank	Δ_{BIC}	decision
$\mathcal{M}_{H1}$	5	0.6504	-1.2965	2	> 10	rejected
$\mathcal{M}_{H2}$	9	1.0280	-2.0481	1	0	optimal

dimension of the parameter space is reduced. No additional parameters need to be introduced for every time point, yielding that the availability of measurements at more time points or more experiments, does not increase the number of parameters. Furthermore, the proposed method provides additional insight into variability within a subpopulation. Intrinsic noise, arising due to the inherent stochasticity of the underlying biological processes, and extrinsic noise, which emerges, for example, from differences in parameters of the cells in a subpopulation, can now be taken into account. Therefore, not only differences between supopulations, but also heterogeneity of the individual subpopulations can be studied. In addition, measurement noise can now be treated separately from the variability of the subpopulations. Moreover, the evolution of the variability of a subpopulation can be predicted, as variances at time points for which no measurements exist can be simulated by the MEs. We validated the method on the example of a conversion process between two species for different scenarios. We showed that the method is able to detect the origin of difference between subpopulations and the existence of additional parameter variability. The confidence intervals using MEs are narrower for the case of measurements at less time points. We expect even more confidence in the estimates by using ODE-MMs with MEs for other processes, in which more information is carried in the variances.

Table 3.5: Model selection for data of Scenario 3 that includes high parameter variability. Both criteria, AIC and BIC, select the true model $\mathcal{M}_{\mathrm{H2}}$, which allows
parameters to vary between individual cells

	n_{θ}	$\log L(\boldsymbol{\theta}^{\mathrm{ML}})(10^4)$	AIC (10^4)	rank	Δ_{AIC}	decision
$\mathcal{M}_{\mathrm{H1}}$	5	-0.4511	0.9032	2	> 10	rejected
$\mathcal{M}_{\mathrm{H2}}$	9	0.8392	-1.6766	1	0	optimal

	n_{θ}	$\log L(\boldsymbol{\theta}^{\mathrm{ML}})(10^4)$	BIC (10^4)	rank	Δ_{BIC}	decision
$\mathcal{M}_{\mathrm{H1}}$	5	-0.4511	0.9065	2	> 10	rejected
$\mathcal{M}_{\mathrm{H2}}$	9	0.8392	-1.6706	1	0	optimal

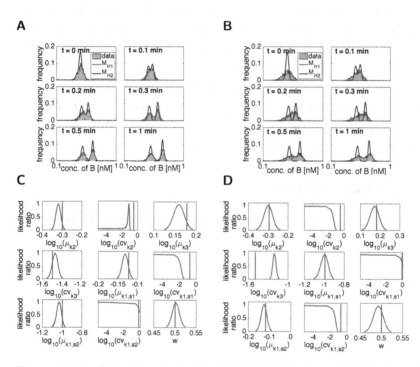

Figure 3.9: Results for Scenario 2 and 3. (**A**) Fitted data for Scenario 2 for both models $\mathcal{M}_{H1}$, allowing no additional parameter variability and $\mathcal{M}_{H2}$, which considers cell-to-cell variability in the kinetic parameters. (**B**) Fitted data for Scenario 3 for $\mathcal{M}_{H1}$ and $\mathcal{M}_{H2}$. The true model $\mathcal{M}_{H2}$ has been selected correctly by AIC and BIC for both scenarios. (**C, D**) Profiles corresponding to the optimal model for Scenario 2 (**C**) and Scenario 3 (**D**). The true values of the parameters, which have been used to generate the data, are indicated by a green line.

3.4 Simultaneous Analysis of Multivariate Measurements

In the previous section we presented ODE-MMs with MEs, which are able to capture variability within a subpopulation. However, correlations between the measurements can still not be detected and taken into account (Problem 4). By just considering the univariate measurements, corresponding to the marginal distributions in Figure 3.10, correlations between the measurands cannot be detected. Figure 3.10A shows positive correlation between measurand A and B, while Figure 3.10B shows negative correlation between the measurements. In the following, we propose ODE-MMs, which can simultaneously analyze multivariate measurements and therefore account for correlated behavior. Additionally, we demonstrate and tackle the numerically instability of the likelihood calculation arising due to the use of mixture probabilities (Problem 5).

3.4.1 Likelihood Function

The likelihood function for ODE-MMs based on multivariate measurement data is given in a general form by

$$L\left(\boldsymbol{\theta}\right) = \prod_{e,k,j} \sum_{s=1}^{n_s} w_s^e \, p\left(\bar{\mathbf{y}}_j^{e,k} | \boldsymbol{\varphi}_s^e\left(t_k\right)\right) \tag{3.5}$$

$$\text{with} \quad \dot{\mathbf{x}}_s^e = f\left(\mathbf{x}_s^e, \boldsymbol{\psi}_s^e, u^e\right), \quad \mathbf{x}_s^e(0) = \mathbf{x}_0\left(\boldsymbol{\psi}_s^e, u^e\right),$$

$$\boldsymbol{\varphi}_s^e = h\left(\mathbf{x}_s^e, \boldsymbol{\psi}_s^e, u^e\right).$$

A   **B**

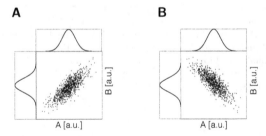

Figure 3.10: Scatterplot and marginals of measurands A and B with (**A**) positive correlation and (**B**) negative correlation. The measurements are given in arbitrary units.

For a simpler notation, we further neglect the indices e, k and j, corresponding to the experiment, time point and single-cell, respectively. While (3.1) and (3.2) consider a univariate measurement $\bar{y} \in \mathbb{R}$, (3.5) gives the probability to observe a multivariate vector $\bar{\mathbf{y}} \in \mathbb{R}^d$. The n_s subpopulations with parameters $\boldsymbol{\psi}_s$ and weights w_s are simulated with an external stimulus u. The multivariate mixture distribution is defined by the parameters $\boldsymbol{\varphi}_s$. Function h links the state vector $\mathbf{x}_s$ of the moments of the species of the system, for which the time evolution is described by f, to the mixture parameters.

For simplicity, we further neglect the subpopulation index s. We use second order moments to describe the mechanisms of the system as in Section 3.3. The measurement is possibly affected by some measurement noise e.g. additive normally distributed measurement noise

$$\bar{\mathbf{y}} = \mathbf{y} + \boldsymbol{\epsilon}, \quad \boldsymbol{\epsilon} \sim \mathcal{N}\left(0, \boldsymbol{\Gamma}\right), \tag{3.6}$$

or multiplicative log-normally distributed measurement noise

$$\bar{\mathbf{y}} = \mathbf{y}\boldsymbol{\epsilon}, \quad \boldsymbol{\epsilon} \sim \log\mathcal{N}\left(0, \boldsymbol{\Gamma}\right). \tag{3.7}$$

The entries of $\boldsymbol{\Gamma}$ are incorporated in the parameter vector. The moments of the measurand without measurement noise are denoted by $\mathbf{m_y} = (m_{\mathbf{y},1}, \ldots, m_{\mathbf{y},d})$ and $\mathbf{C_y}$ and can be calculated from $\mathbf{x}$.

3.4.2 Multivariate Mixture of Normal and Log-Normal Distributions

In this thesis we focus on the mixture of d-dimensional multivariate normal distributions

$$\mathcal{N}(\mathbf{y}|\boldsymbol{\mu}, \boldsymbol{\Sigma}) = \frac{1}{(2\pi)^{\frac{d}{2}} \det(\boldsymbol{\Sigma})^{\frac{1}{2}}} e^{-\frac{1}{2}(\mathbf{y}-\boldsymbol{\mu})^T \boldsymbol{\Sigma}^{-1}(\mathbf{y}-\boldsymbol{\mu})},$$

and multivariate log-normal distributions

$$\log\mathcal{N}(\mathbf{y}|\boldsymbol{\mu}, \boldsymbol{\Sigma}) = \frac{1}{(2\pi)^{\frac{d}{2}} \det(\boldsymbol{\Sigma})^{\frac{1}{2}} \left(\prod_{i=1}^{d} y_i\right)} e^{-\frac{1}{2}(\log(\mathbf{y})-\boldsymbol{\mu})^T \boldsymbol{\Sigma}^{-1}(\log(\mathbf{y})-\boldsymbol{\mu})}.$$

with mixture parameters $\boldsymbol{\varphi} = (\boldsymbol{\mu}, \boldsymbol{\Sigma})$.

Gradient of Multivariate Normal and Log-Normal Distribution

The gradient of a multivariate normal distribution $\mathcal{N}(\mathbf{y}|\boldsymbol{\mu}(\theta), \boldsymbol{\Sigma}(\theta))$ with respect to θ is known to be

$$
\frac{\partial}{\partial \theta} \mathcal{N}(\mathbf{y}|\boldsymbol{\mu}, \boldsymbol{\Sigma}) = -\frac{1}{2} \mathcal{N}(\mathbf{y}|\boldsymbol{\mu}, \boldsymbol{\Sigma}) \cdot \left(\mathrm{Tr}\left(\boldsymbol{\Sigma}^{-1} \frac{\partial \boldsymbol{\Sigma}}{\partial \theta} \right) + (\boldsymbol{\mu} - \mathbf{y})^T \boldsymbol{\Sigma}^{-1} \left(\frac{\partial \boldsymbol{\mu}}{\partial \theta} \right)^T \right.
$$
$$
\left. + \left(\frac{\partial \boldsymbol{\mu}}{\partial \theta} \right)^T \boldsymbol{\Sigma}^{-1} (\boldsymbol{\mu} - \mathbf{y}) + (\boldsymbol{\mu} - \mathbf{y})^T \frac{\partial \boldsymbol{\Sigma}^{-1}}{\partial \theta} (\boldsymbol{\mu} - \mathbf{y}) \right).
$$

For simplicity we write $\boldsymbol{\mu} = \boldsymbol{\mu}(\theta)$ and $\boldsymbol{\Sigma} = \boldsymbol{\Sigma}(\theta)$. The derivative for the multivariate log-normal distribution can be obtained using the relation

$$
\log\mathcal{N}(\mathbf{y}|\boldsymbol{\mu}, \boldsymbol{\Sigma}) = \mathcal{N}(\log(\mathbf{y})|\boldsymbol{\mu}, \boldsymbol{\Sigma}) \left(\prod_{i=1}^{d} y_i \right)^{-1}.
$$

To simplify and speed up the calculation of the gradient we use the identity

$$
\mathrm{Tr}\left(\mathbf{A}^T \mathbf{B} \right) = \sum_{i,j} (\mathbf{A} \circ \mathbf{B}),
$$

with $(\mathbf{A} \circ \mathbf{B})_{ij} = (A_{ij} B_{ij})$ being the Hadamard product.

Mixture of Multivariate Normal Distributions

We now present the calculations needed to constrain the mean and covariance of a multivariate normal distribution by MEs. Given the means $\mathbf{m}_y$, covariances $\mathbf{C}_y$ and the corresponding sensitivities $\frac{\partial \mathbf{m}_y}{\partial \theta}$ and $\frac{\partial \mathbf{C}_y}{\partial \theta}$, we can directly define $\boldsymbol{\mu} = \mathbf{m}_y$, $\boldsymbol{\Sigma} = \mathbf{C}_y$ and the derivatives

$$
\frac{\partial \boldsymbol{\mu}}{\partial \theta} = \frac{\partial \mathbf{m}_y}{\partial \theta}, \text{ and } \left(\frac{\partial \boldsymbol{\Sigma}}{\partial \theta} \right)_{ij} = \left(\frac{\partial \mathbf{C}_y}{\partial \theta} \right)_{ij} = \frac{\partial C_{y,ij}}{\partial \theta}.
$$

The multivariate normal distributions can e.g. be used together with additive normal measurement noise $\boldsymbol{\Sigma} = \mathbf{C}_y + \boldsymbol{\Gamma}$ as in (3.6). An example for assuming independent measurement noise is given by

$$\mathbf{\Gamma} = \begin{pmatrix} \sigma_{\epsilon,1}^2 & 0 & 0 \\ 0 & \ddots & 0 \\ 0 & 0 & \sigma_{\epsilon,d}^2 \end{pmatrix} .$$

Other considerations, such as correlated behavior of the measurement noise for the different measurements, can also be included.

Mixture of Multivariate Log-Normal Distributions

For the log-normal distribution we use two different parametrizations. If the mean obtained by the MEs describes the mean of the log-normal distribution, we use

$$m_{\mathbf{y},i} = e^{\mu_i + \frac{1}{2}\Sigma_{ii}},$$
$$C_{\mathbf{y},ij} = e^{\mu_i + \mu_j + \frac{1}{2}(\Sigma_{ii} + \Sigma_{jj})}(e^{\Sigma_{ij}} - 1),$$

for $i, j = 1, \ldots, d$. This yields the mixture parameters

$$\mu_i = \log(m_{\mathbf{y},i}) - \frac{1}{2}\Sigma_{ii},$$
$$\Sigma_{ij} = \log(\frac{C_{\mathbf{y},ij}}{m_{\mathbf{y},i} m_{\mathbf{y},j}} + 1),$$

and their derivatives

$$\frac{\partial \mu_i}{\partial \theta} = \frac{1}{m_{\mathbf{y},i}} \frac{\partial m_{\mathbf{y},i}}{\partial \theta} - \frac{1}{2} \frac{\partial \Sigma_{ii}}{\partial \theta},$$

$$\frac{\partial \Sigma_{ij}}{\partial \theta} = \frac{1}{\frac{C_{\mathbf{y},ij}}{m_{\mathbf{y},i} m_{\mathbf{y},j}} + 1} \frac{\partial \left(\frac{C_{\mathbf{y},ij}}{m_{\mathbf{y},i} m_{\mathbf{y},j}}\right)}{\partial \theta}$$

$$= \frac{1}{\frac{C_{\mathbf{y},ij}}{m_{\mathbf{y},i} m_{\mathbf{y},j}} + 1} \frac{m_{\mathbf{y},i} m_{\mathbf{y},j} \frac{\partial C_{\mathbf{y},ij}}{\partial \theta} - C_{\mathbf{y},ij}(m_{\mathbf{y},i} \frac{\partial m_{\mathbf{y},j}}{\partial \theta} + m_{\mathbf{y},j} \frac{\partial m_{\mathbf{y},i}}{\partial \theta})}{(m_{\mathbf{y},i} m_{\mathbf{y},j})^2} .$$

Another parametrization is given by assuming that the mean obtained by the MEs describes the median of the log-normal distribution. This distribution assumption is combined with log-normally distributed measurement noise (3.7).

3.4.3 Robust Computation of Mixture Probabilities

When using mixture models, the calculation of the likelihood is generally unstable (Problem 5). We explain and address this problem in the following. We are interested in the likelihood function of a mixture model with n_s mixture components given by

$$p(\mathbf{y}|\boldsymbol{\theta}) = \sum_{s=1}^{n_s} w_s p(\mathbf{y}|\boldsymbol{\varphi}_s), \text{ with } \boldsymbol{\theta} = \{(w_s, \boldsymbol{\varphi}_s)\}_{s=1}^{n_s}.$$

For simplicity we denote $p(\mathbf{y}|\boldsymbol{\theta})$ by p and the probabilities $p(\mathbf{y}|\boldsymbol{\varphi}_s)$ for the individual components by p_s. The goal is to calculate the log-likelihood

$$\log(p) = \log\left(\sum_{s=1}^{n_s} w_s p_s\right)$$

in a numerically stable way, which can be achieved by the following reformulation. Let $q_s = \log(p_s)$ for $s = 1\ldots,n_s$, then

$$s_{\max} = \arg\max_s q_s,$$

$$\log(p) = \log\left(\sum_{s=1}^{n_s} w_s e^{q_s}\right)$$

$$= \log\left(\frac{(\sum_{s=1}^{n_s} w_s e^{q_s}) w_{s_{\max}} e^{q_{s_{\max}}}}{w_{s_{\max}} e^{q_{s_{\max}}}}\right)$$

$$= \log\left(\frac{\sum_{s=1}^{n_s} w_s e^{q_s}}{w_{s_{\max}} e^{q_{s_{\max}}}}\right) + \log\left(w_{s_{\max}} e^{q_{s_{\max}}}\right)$$

$$= \log\left(1 + \sum_{s \neq s_{\max}} \frac{w_s}{w_{s_{\max}}} \left(e^{q_s - q_{s_{\max}}}\right)\right) + \log(w_{s_{\max}}) + q_{s_{\max}}.$$

As $0 \leq e^{q_s - q_{s_{\max}}} \leq 1$ the reformulation gives better numerical properties than the direct computation of $\log(p)$, which gets unstable for p close to 0. For the calculation of the gradient, we have

$$\frac{d\log(p)}{d\theta} = \frac{1}{p}\frac{dp}{d\theta} = \sum_{s=1}^{n_s} \frac{p_s}{\sum_{j=1}^{n_s} w_j p_j} H_s,$$

with H_s such that

$$p_s H_s = \frac{dw_s p_s}{d\theta} = p_s \frac{dw_s}{d\theta} + w_s \frac{dp_s}{d\theta} \,. \tag{3.8}$$

This is again unstable for p close to 0 and needs to be reformulated. Using

$$\frac{p_s}{\sum_{j=1}^{n_s} w_j p_j} = \frac{\frac{p_s}{p_{s_{max}}}}{\sum_{j=1}^{n_s} w_j \frac{p_j}{p_{s_{max}}}} = \frac{e^{q_s - q_{s_{max}}}}{\sum_{j=1}^{n_s} w_j e^{q_j - q_{s_{max}}}} \,,$$

we obtain

$$\frac{d \log(p)}{d\theta} = \frac{1}{\sum_{j=1}^{n_s} w_j e^{q_j - q_{s_{max}}}} \sum_{s=1}^{n_s} e^{q_s - q_{s_{max}}} H_s \,. \tag{3.9}$$

Example (Normal Distribution).   For a normal distribution $\mathcal{N}(\mu_s, \sigma_s^2)$ with parameters $\varphi_s = (\mu_s, \sigma_s)$ and w_s that depend on θ, we calculate

$$q_s(y|\varphi_s) = -\frac{1}{2} \left(\frac{y - \mu_s}{\sigma_s} \right)^2 - \log(\sqrt{2\pi}) - \log(\sigma_s) \,.$$

The term H_s defined in (3.8) is

$$H_s = \frac{dw_s}{d\theta} + \frac{w_s}{\sigma_s} \left(\frac{y - \mu_s}{\sigma_s} \frac{d\mu_s}{d\theta} + \left(\left(\frac{y - \mu_s}{\sigma_s} \right)^2 - 1 \right) \frac{d\sigma_s}{d\theta} \right) \,.$$

and can be used for the stable calculation of the gradient according to (3.9).

Comparison of Calculations of The Log-Likelihood Function

To evaluate the recalculation, we performed the parameter estimation in the next section for the multivariate case with and without the reformulation above. Both procedures find the same optimum. While the optimizer has a convergence rate of 0.71 using the robust calculation, the convergence rate without the reformulation is 0.01. The corresponding log-likelihood values are shown in Figure 3.11.

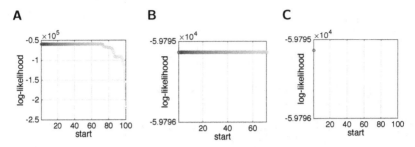

Figure 3.11: Performance comparison for the numerically more stable and the classical calculation of the likelihood. The red circles mark the value of the log-likelihood function corresponding to the ML estimate. (**A**, **B**) show the optimizer output for the robust calculation. (**A**) All (local) optima found by the optimizer. (**B**) The best log-likelihood value is found in 71 out of 100 runs. The difference of the values is below 10^{-10}. (**C**) Only 1 initial value of 100 has a probabilty greater than 0 using the classical calculation.

3.4.4 Simulation Example: Conversion Reaction

For the validation of the method we generate data of a conversion process as in Section 3.3.2, comprising the conversion between two species A and B.

Artificial Data

As we want to analyze multivariate data, we measure both species A and B. We generate trajectories of 1000 cells in some volume $\Omega = 1000$. The total number of molecules $N_0 = A + B$ is log-normally distributed with mean $\mu_{N_0} = 1000$ and variance $\sigma_{N_0}^2 = 2500$. The cell population has a subpopulation structure due to differences in the response to stimulus u. The parameter k_1, describing the stimulus dependent conversion of A to B, is set to $k_{1,s1} = 0.1$ for the low responsive subpopulation and to $k_{1,s2} = 0.75$ for the high responsive subpopulation. Both subpopulations have the same size ($w = 0.5$). The parameters shared by the subpopulations are given by $k_2 = 0.5$ and $k_3 = 1.5$. Measurements **y** of the absolute numbers of A and B are assumed to be recorded at $t = 0, 0.1, 0.2, 0.3, 0.5$ and 1 minutes. The parameter vector that needs to be estimated from the data is given by $\boldsymbol{\theta} = (k_{1,s1}, k_{1,s2}, k_2, k_3, \mu_{N_0}, \sigma_{N_0}, w)^T$.

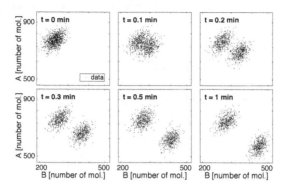

Figure 3.12: Artificial multivariate data of a conversion process of A and B with cell-to-cell variability in the total number of molecules N_0.

Moment Equations for a Conversion Process with Variability in N_0

To obtain the MEs, we assume to have three species A, B and $N_0 = A + B$. The number of molecules for species A and B changes according to the reactions of the conversion process introduced in Section 3.3.2. The total number of molecules N_0 is assumed to be distributed with some mean μ_{N_0} and variance σ_{N_0} across the cells. Within an individual cell, the number of molecules is constant. This yields the MEs (Kazeroonian *et al.*, 2016)

$$\frac{dm_{\mathrm{B}}}{dt} = (k_1 + k_3)m_{\mathrm{A}} - k_2 m_{\mathrm{B}},$$

$$\frac{dm_{\mathrm{A}}}{dt} = k_2 m_{\mathrm{B}} - (k_1 + k_3)m_{\mathrm{A}},$$

$$\frac{dm_{N_0}}{dt} = 0,$$

$$\frac{dC_{\mathrm{B,B}}}{dt} = (k_1 + k_3)m_{\mathrm{B}} + k_2 m_{\mathrm{A}} + 2(k_1 + k_3)C_{\mathrm{B,A}} - 2k_2 C_{\mathrm{B,B}},$$

$$\frac{dC_{\mathrm{B,A}}}{dt} = -(k_1 + k_3)m_{\mathrm{B}} - k_2 m_{\mathrm{A}} - (k_1 + k_2 + k_3)C_{\mathrm{B,A}} + k_1 C_{\mathrm{A,A}} + k_2 C_{\mathrm{B,B}},$$

$$\frac{C_{\mathrm{B},N_0}}{dt} = (k_1 + k_3)C_{A,N_0} - k_2 C_{\mathrm{B},N_0},$$

$$\frac{C_{\mathrm{A,A}}}{dt} = (k_1 + k_3)m_{\mathrm{A}} + k_2 m_{\mathrm{B}} + 2k_2 C_{\mathrm{B,A}} - 2(k_1 + k_3)C_{\mathrm{A,A}},$$

$$\frac{C_{\mathrm{A},N_0}}{dt} = -(k_1 + k_3)C_{A,N_0} + k_2 C_{\mathrm{B},N_0},$$

$$\frac{C_{N_0,N_0}}{dt} = 0.$$

Using conservation of mass and calculation rules for covariances we obtain the following initial conditions

$$m_B(0) = \frac{k_2}{k_2 + k_3}\mu_{N_0},$$

$$m_A(0) = \left(1 - \frac{k_2}{k_2 + k_3}\right)\mu_{N_0},$$

$$m_{N_0}(0) = \mu_{N_0},$$

$$C_{A,A}(0) = \sigma_{N_0}^2 - 2C_{B,N_0} + C_{B,B}(0),$$

$$C_{B,B}(0) = \frac{k_2 m_A(0) + k_3 m_B(0) + 2k_2 C_{B,N_0}}{2(k_2 + k_3)},$$

$$C_{A,N_0}(0) = \frac{k_2}{k_2 + k_3}\sigma_{N_0}^2,$$

$$C_{B,N_0}(0) = \frac{k_3}{k_2 + k_3}\sigma_{N_0}^2,$$

$$C_{N_0,N_0} = \sigma_{N_0}^2.$$

Comparison of Multivariate and Univariate Analysis of Measurements

To analyze how much information is gained by considering multivariate measurements, we perform parameter estimation first by simultaneously analyzing both measurements and then by analyzing one measurement at a time. The parameter restrictions are the same for both cases, namely the kinetic parameters and mean number of molecules μ_{N_0} are assumed to lie in $[10^{-6}, 10^4]$, the weight within $[0, 1]$ and the coefficient of variation for the number of molecules $cv_{N_0} = \sigma_{N_0}/\mu_{N_0}$ within $[10^{-6}, 10^4]$. For the univariate case we analyze the measurements of A and B independently, while we simultaneously use both measurements in the likelihood (3.5) for the multivariate case. The fits corresponding to the best parameters of 100 multi-start local optimization runs are depicted in Figures 3.13A-C. The model can explain the data quite well. The comparison of the likelihoods in Figure 3.13D shows that information is gained by using both measurements at the same time.

We presented a method, which considers multivariate measurements simultaneously to infer parameters. This allows to capture correlated behavior between the measurements, which would not be possible if the measurements are considered separately. More information can be extracted from the data yielding higher confidence in the estimates.

Furthermore, we proposed a reformulation of the likelihood of mixture probabilities to obtain a robust computation, which cannot only be used within ODE-MM, but can be applied to other approaches that require the calculation of mixture probabilities.

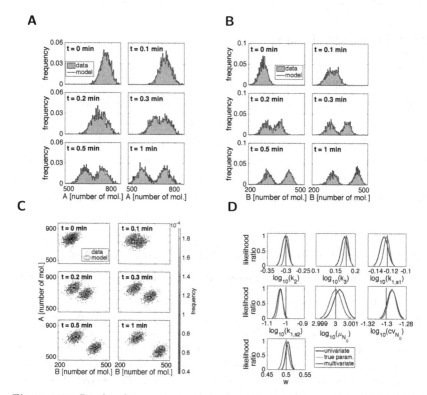

Figure 3.13: Results of parameter estimation and identifiability analysis. (**A**, **B**) Fitted data for species A and B using the univariate implementation of ODE-MMs. (**C**) Fitted multivariate measurements of A and B. The level sets of the multivariate normal mixture distribution are visualized. (**D**) Comparison of the profile likelihoods for both cases. The confidence intervals are wider if the measurements are considered seperately.

3.5 Application Example: NGF-Induced Erk Signaling

In the previous sections, we implemented a method based on ODE-MMs, which is able to capture variability within a subpopulation and analyze multivariate measurements. We apply our method to data of NGF-induced Erk signaling in primary sensory neurons, a process that has already been explained and studied in Section 3.2.1.

3.5.1 Experimental Data

We investigate three experiments, comprising kinetics and dose responses. For experiment 1 and 2, only the pErk concentration has been measured (see Figures 3.14A and B). For experiment 3, multivariate measurements of pErk and total Erk levels are available (see Figure 3.14C). For details of the collection of the data we refer to (Hasenauer *et al.*, 2014b), in which the first two experiments have been analyzed with ODE-MMs using RREs. As this method is not able to analyze multivariate data simultaneously, the two-dimensional measurements have not yet been included into the parameter estimation procedure and have only been used to validate the results obtained based on the univariate measurements.

3.5.2 Moment Equations for The Simple Pathway Model of NGF-Induced Erk Signaling with Variability in Total Erk Levels

Since the data visualized in Figure 3.14C shows that the total Erk levels vary between individual cells, our model assumes total Erk levels to be distributed with some mean and variance. As we use MEs that describe the evolution of the concentrations, we assume a volume of $\Omega = 500\mu m^3$ for a neuron. We scale the system with $\Omega Av = 301.1$, with Av being Avogadros constant. The concentration, which is obtained by dividing the number of molecules by ΩAv, is given in units of nM. The concentration of the complex TrkA:NGF is denoted by C, [pErk] by P, $[Erk]_0$ by E_0 and $[TrkA]_0$ by T_0. The system is stimulated with $[NGF]_0$ denoted by NGF_0, corresponding to u in (3.5). Given the parameters $k_1, k_2, k_3, k_4, k_5, T_0, \mu_{E_0}$ and σ_{E_0}, we obtain the following MEs using the toolbox CERENA (Kazeroonian *et al.*, 2016):

$$\frac{dm_C}{dt} = -k_2 m_C + k_1 \frac{C_{C,C} + (\Omega Av NGF_0 - m_C)(T_0 - m_C)}{\Omega Av},$$

$$\frac{dm_P}{dt} = -k_5 m_P - k_4(m_P - m_{E_0}) + k_3 \frac{C_{C,E_0} - C_{C,P} - m_C(m_P - m_{E_0})}{\Omega Av},$$

$$\frac{dm_{E_0}}{dt} = 0,$$

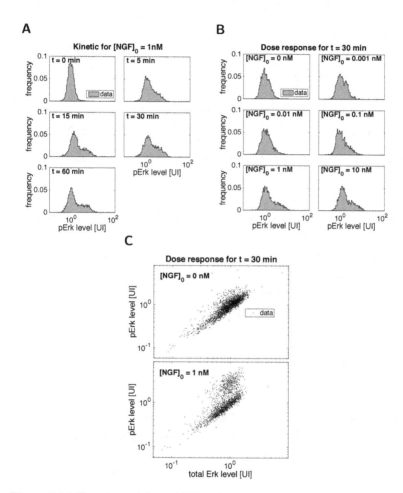

Figure 3.14: Experimental data of NGF-induced Erk phosphorylation. (**A**) Univariate kinetic measurements of pErk for 18797 cells. (**B**) Univariate dose response measurements of pErk for 12205 cells. (**C**) Multivariate dose response measurements of pErk and total concentration of Erk for 4134 cells.

$$\frac{dC_{C,C}}{dt} = k_2(m_C - 2C_{C,C}) + k_1\frac{C_{C,C} + (\Omega A v NGF_0 - m_C)(T_0 - m_C)}{\Omega A v} -$$
$$\frac{2C_{C,C}(\Omega A v NGF_0 + T_0 - 2m_C)}{\Omega A v},$$

$$\frac{dC_{C,P}}{dt} = k_4(C_{C,E_0} - C_{C,P}) - k_5 C_{C,P} - k_2 C_{C,P} +$$
$$k_3\frac{C_{C,E_0}m_C - C_{C,P}m_C - C_{C,C}(m_P - m_{E_0})}{\Omega A v} - k_1\frac{C_{C,P}k_1(\Omega A v NGF_0 + T_0 - 2m_C)}{\Omega A v},$$

$$\frac{dC_{C,E_0}}{dt} = -k_2 C_{C,E_0} - k_1\frac{C_{C,E_0}(\Omega A v NGF_0 + T_0 - 2m_C)}{\Omega A v},$$

$$\frac{dC_{P,P}}{dt} = k_4(2C_{P,E_0} - 2C_{P,P} - (m_P - m_{E_0})) + k_5(m_P - 2C_{P,P}) +$$
$$k_3\frac{C_{C,E_0} - C_{C,P} - 2C_{P,P}m_C + 2C_{P,E_0}m_C - 2C_{C,P}(m_P - m_{E_0}) - m_C(m_P - m_{E_0})}{\Omega A v},$$

$$\frac{dC_{P,E_0}}{dt} = k_4(C_{E_0 E_0} - C_{P,E_0}) - k_5 C_{P,E_0} + k_3\frac{C_{E_0,E_0}m_C - C_{P,E_0}m_C - C_{C,E_0}(m_P - m_{E_0})}{\Omega A v},$$

$$\frac{dC_{E_0,E_0}}{dt} = 0.$$

The initial condition is the steady state of the system without stimulus, i.e., $NGF_0 = 0$,

$$m_P(0) = \frac{k_4}{k_4 + k_5}\mu_{E_0},$$

$$m_{E_0}(0) = \mu_{E_0},$$

$$C_{P,P}(0) = \frac{2k_4^2\sigma_{E_0}^2 + 2k_4 k_5\mu_{E_0}}{2(k_4 + k_5)^2},$$

$$C_{P,E_0}(0) = \frac{k_4}{k_4 + k_5}\sigma_{E_0}^2,$$

$$C_{E_0,E_0}(0) = \sigma_{E_0}^2,$$

$$m_C(0) = 0, \; C_{C,C}(0) = 0, \; C_{C,P}(0) = 0, \; C_{C,E_0}(0) = 0.$$

The measurements y^e for the univariate kinetic ($e = 1$) and dose response ($e = 2$) data, and the multivariate dose response data ($e = 3$) are then given in concentrations by

$$y^e = s_1 P = s_1[pErk], \qquad\qquad \text{for } e = 1, 2,$$

$$y^e = \begin{pmatrix} s_2 P + b \\ s_3 E_0 \end{pmatrix} = \begin{pmatrix} s_2[pErk] + b \\ s_3[Erk]_0 \end{pmatrix}, \qquad\qquad \text{for } e = 3,$$

with additional experiment-specific scaling parameters s_1, s_2, s_3 and offset parameter denoted by b.

Table 3.6: Results for model selection for NGF-induced Erk signaling.

	n_s	n_θ	$\log L(\boldsymbol{\theta}^{\mathrm{ML}})(10^4)$	AIC (10^4)	Δ_{AIC}	BIC (10^4)	Δ_{BIC}	rank	decision
$\mathcal{M}_{\mathrm{H1}}$	1	13	-4.1263	8.2252	>10	8.2662	>10	3	rejected
$\mathcal{M}_{\mathrm{H2}}$	2	15	-3.9135	7.8300	>10	7.8427	>10	2	rejected
$\mathcal{M}_{\mathrm{H3}}$	2	16	-3.9079	7.8190	0	7.8325	0	1	optimal

3.5.3 Hypothesis Testing

To assess our method for real experimental data, we test the following hypotheses:

H1 No difference between subpopulations.

H2 Different levels of TrkA (TrkA$_{0,i}$ for subpopulation i).

H3 Different levels of TrkA and different mean concentrations of Erk (TrkA$_{0,i}$ and $\mu_{\mathrm{E}_{0,i}}$ for subpopulation i).

As the log-normal distribution parametrized by its median has been selected as the optimal distribution by Hasenauer *et al.* (2014b) based on the univariate data, we use a multivariate log-normal distribution as mixture distribution and constrain the median by the MEs. H1 includes variability in total Erk levels, but assumes that the cell population consist of only one population. H2 and H3 consider a subpopulation structure, emerging due to different levels of TrkA. For H3, also the parameter $\mu_{\mathrm{E}_{0,i}}$ describing the mean of total Erk levels differs between the subpopulations. We parametrize the variance $\sigma^2_{\mathrm{E}_{0,i}}$ by the coefficient of variation $cv_{\mathrm{E}_{0,i}}$. In addition, we assume independent log-normally distributed multiplicative measurement noise. All parameters besides the subpopulation weight, which is assumed to lie within $[0,1]$, are restricted to the interval $[10^{-10}, 10^{10}]$.

The results for parameter estimation with 1000 multi-starts and model selection using AIC and BIC are shown in Figure 3.15 and Table 3.6. The fits indicate that the model can explain the data. Both criteria select $\mathcal{M}_{\mathrm{H3}}$, indicating that the origin of difference are not only TrkA levels, but also different means of total levels Erk.

We applied ODE-MMs to kinetic and dose response experiments of NGF-induced Erk signaling, comprising different dimensions of measurements. Using not only the measurements of pErk, but also the two-dimensional measurements of pErk and total Erk levels gives a more detailed insight into the underlying system. By performing parameter estimation and model selection with our proposed method, it was possible to account for variability in total Erk levels and to identify a further difference between the subpopulations, namely the mean of total Erk levels.

3.6 Discussion and Outlook

In this chapter, we used and extended ODE-MMs to study subpopulation structures and dynamics. Based on novel data for NGF-induced Erk signaling, we evaluated the methods applicability to unravel differences in subpopulation response between experimental conditions. A mechanistic description of the subpopulation by MEs yields not only a reduced number of parameters, it also enables us to capture intrinsic and extrinsic variability of a subpopulation. In addition, we enhanced the method to analyze multivariate measurements and therefore account for correlations between the measurands. We successfully tackled numerical instability, which not only arises for ODE-MMs but for other methods using mixture distributions. The application of our method to examples for a conversion process and real data of NGF-induced Erk signaling revealed an improved acquisition of the data.

In this thesis, we used the normal and log-normal distribution to mix the subpopulations. Other distributions, for example the t-distribution and the skew-t distribution (Pyne *et al.*, 2009) can also be used within ODE-MMs. For distributions that are defined by more than their first two moments, higher order MEs can be used to describe the system and be linked to the mixture distributions.

Not only the mixture distribution can be exchanged. Other representations of the system, which also account for variability within the subpopulations can be studied, e.g. the linear noise approximation (Kampen, 2007, Chapter X). If the system comprises species of low- and high-copy numbers, the method of conditional moment equations described in Section 2.2.4 can be incorporated into ODE-MMs.

So far, the number of subpopulations needs to be predefined and can be chosen by performing model selection for different numbers of mixture components . An adaptive approach to choose this number would be a Bayesian version of the method that uses reversible jump Markov chain Monte Carlo techniques (Murphy, 2012). These methods are able to select an appropriate number of subpopulations by performing parameter estimation across parameter spaces of different dimensions.

In summary, our results prove that ODE-MMs are a flexible tool for the analysis of heterogeneous populations. Our extensions facilitate to extract more information from the data and therefore provide an improved insight into the mechanisms of the biological processes.

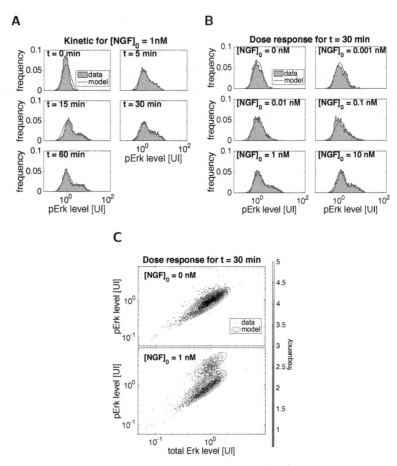

Figure 3.15: Fitted data of NGF-induced Erk phosphorylation for model $\mathcal{M}_{H3}$. (**A**) Univariate fit of kinetic measurements of pErk. (**B**) Univariate fit of dose response measurements of pErk. (**C**) Multivariate fit for dose response measurements of pErk and total concentration of Erk. The model is represented by level sets of the frequency, given by a multivariate log-normal distribution.

4 Approximate Bayesian Computation for Single-Cell Time-Lapse Data Using Multivariate Statistics

In the previous chapter we focused on modeling and parameter estimation for single-cell snapshot data. We provided and assessed a method, which not only considers the mean behavior of the cell population, but also the second order moments. In this chapter we go one step further in the direction of treating cells as individuals by modeling single-cell time-lapse data with continuous time Markov chains (CTMCs) (Gillespie, 2007).

In Section 4.1, we formulate the key problems that arise in parameter inference for single-cell time-lapse data and that are addressed in this chapter. Section 4.2 introduces approximate Bayesian computation (ABC), a method used for parameter estimation without the usage of a likelihood function. An essential step of ABC is the comparison of two data sets. Therefore, we present in Section 4.3 two multivariate test statistics that are suitable for the task of comparing time-series. We study how these statistics can be incorporated into the ABC scheme. Different approaches are evaluated on a bivariate normal distribution. In Section 4.4, our method is applied to artificial time-series of a one-stage model of gene expression. Here, we assess our method for equilibrium and non-equilibrium data as well as for data including parameter variability and tree-structured data. In Section 4.5, we summarize and discuss our results.

4.1 Introduction and Problem Statement

For small molecule numbers stochastic effects may complicate the analysis on the population level. Thus, the system should be modeled in a stochastic way accounting for the randomness of the underlying processes (Wilkinson, 2009). CTMCs model the births and deaths of single molecules and therefore capture intrinsic noise.

To estimate parameters of a CTMC, likelihood-based methods can be applied. These methods consider all possible paths of the stochastic process by evaluating the transition density, e.g. using the finite state projection (FSP) (Munsky & Khammash, 2006). As this is computationally demanding and merely tractable for simple processes and moderate system sizes, likelihood-free approaches have been developed, which are also called approximate Bayesian computation (ABC) methods (Marin et al., 2012). ABC methods circumvent the evaluation of the likelihood. A parameter is accepted if the distance between simulated and measured data is sufficiently small, and rejected otherwise. The performance and convergence of ABC depends crucially on the employed distance measure.

Single-cell time-lapse data are highly multivariate, especially if measurements at several time points exist. While it is possible to collect thousands of measurements e.g. flow cytometry, the generation of single-cell time-series is time-consuming, as its requires several steps and the cell tracking often has to be done manually. On the other hand, information is gained as cells are tracked over time, and this source of information should be taken into account in the parameter estimation procedure. When inferring the parameters for single-cell time-lapse data, the following problems arise:

Problem 1 Single-cell time-series are multivariate and appropriate distance measures are not yet available.

Problem 2 The sample size of single-cell time-series is small.

In this thesis, we will take a population perspective rather than fitting trajectories individually (Dargatz, 2010). Trajectories of individual cells are samples from a high-dimensional path distribution. Accordingly, distance measures for ABC are provided by multivariate test statistics.

4.2 Extended Introduction to Approximate Bayesian Computation

Approximate Bayesian computation (ABC) has first been introduced by Pritchard *et al.* (1999). The method has been extended and improved continuously (see e.g. (Marin *et al.*, 2012)) and covers a broad range of applications (see (Csilléry *et al.*, 2010) for an overview).

4.2.1 Algorithms

The most basic ABC algorithm is the ABC rejection sampler, which is explained in Algorithm 4.1. The goal is to obtain samples from the posterior distribution $p(\boldsymbol{\theta}|\mathcal{D}^{\text{obs}})$ of the parameters based on observed data $\mathcal{D}^{\text{obs}}$. Therefore, a data set $\mathcal{D}^{\text{sim}}$ is simulated with respect to a parameter $\boldsymbol{\theta}$ that has been sampled from the prior distribution $p(\boldsymbol{\theta})$. This data set is then compared to the experimental data set. The comparison is carried out by calculating some distance $d(\mathcal{D}^{\text{obs}}, \mathcal{D}^{\text{sim}})$ between both data sets. If the distance is below a certain predefined threshold ϵ, the sampled parameter is accepted as a sample of the posterior distribution. Otherwise, the sample is rejected and another parameter is sampled from the prior until it produces a distance below the desired threshold. This whole procedure is repeated several times to obtain P samples of the posterior approximation of interest. The approximated posterior then converges to the true posterior for $\epsilon \to 0$.

The main drawback of ABC rejection is that the acceptance rate can be low if prior and posterior distribution are very different. Some extensions tackle this problem by combining ABC with other methods such as Markov chain Monte Carlo (Marjoram *et al.*, 2003) or sequential Monte Carlo methods (Toni *et al.*, 2009). The latter algorithm, ABC with sequential Monte Carlo (ABC SMC) is explained in Algorithm 4.2. The first step is basically an ABC rejection step, in which the particles are sampled from the prior and are accepted if the corresponding simulated data set produces a distance below the first threshold ϵ_1. Every particle obtains a weight $w_i^{(t)}$, which is used to sample the next population. For intermediate populations with an index $t > 2$, the particles are sampled from the previous population and additionally perturbed with respect to the perturbation kernel $K_t(\cdot|\cdot)$. The particles of the final population are then the desired samples of the approximated posterior distribution.

Algorithm 4.1: ABC rejection

Input: Data set $\mathcal{D}^{\mathrm{obs}}$, prior $p(\boldsymbol{\theta})$, threshold ϵ, simulation function $p(\mathcal{D}|\boldsymbol{\theta})$, number
  of particles P.

Result: Samples of $p(\boldsymbol{\theta}|d(\mathcal{D}^{\mathrm{obs}}, \mathcal{D}^{\mathrm{sim}}) \leq \epsilon)$.

for $i = 1, \ldots, P$ **do**

 | **repeat**
 | | Draw $\boldsymbol{\theta}_i^{(1)} \sim p(\boldsymbol{\theta})$ and simulate $\mathcal{D}^{\mathrm{sim}} \sim p(\mathcal{D}|\boldsymbol{\theta}_i)$.
 | **until** $d(\mathcal{D}^{\mathrm{obs}}, \mathcal{D}^{\mathrm{sim}}) < \epsilon$

end

Algorithm 4.2: ABC SMC (Toni *et al.*, 2009)

Input: Data set $\mathcal{D}^{\mathrm{obs}}$, prior $p(\boldsymbol{\theta})$, perturbation kernel $K_t(\cdot|\cdot)$, threshold schedule
  $\epsilon_1 > \ldots > \epsilon_T$ for the T populations, simulation function $p(\mathcal{D}|\boldsymbol{\theta})$, number
  of particles P

Result: Samples of $p(\boldsymbol{\theta}|d(\mathcal{D}^{\mathrm{obs}}, \mathcal{D}^{\mathrm{sim}}) \leq \epsilon)$.

Set population index $t = 1$.

for $i = 1, \ldots, P$ **do**

 | **repeat**
 | | Draw $\boldsymbol{\theta}_i^{(1)} \sim p(\boldsymbol{\theta})$ and simulate $\mathcal{D}^{\mathrm{sim}} \sim p(\mathcal{D}|\boldsymbol{\theta}_i^{(1)})$
 | **until** $d(\mathcal{D}^{\mathrm{obs}}, \mathcal{D}^{\mathrm{sim}}) < \epsilon_1$
 | Set $w_i^{(t)} = 1/P$, the weight of particle $\boldsymbol{\theta}_i^{(t)}$.
 | **for** $t = 2, \ldots, T$ **do**
 | | **for** $i = 1, \ldots, P$ **do**
 | | | **repeat**
 | | | | Pick $\boldsymbol{\theta}_i^*$ from the $\boldsymbol{\theta}_j^{(t-1)}$'s with probabilities $w_j^{(t-1)}$.
 | | | **until** $p(\boldsymbol{\theta}_i^*) > 0$
 | | | **repeat**
 | | | | Draw $\boldsymbol{\theta}_i^{(t)} \sim K_t(\boldsymbol{\theta}_i^{(t)}|\boldsymbol{\theta}_i^*)$ and simulate $\mathcal{D}^{\mathrm{sim}} \sim p(\mathcal{D}|\boldsymbol{\theta}_i^{(t)})$.
 | | | **until** $d(\mathcal{D}^{\mathrm{obs}}, \mathcal{D}^{\mathrm{sim}}) < \epsilon_t$
 | | | Compute new weights as $w_i^{(t)} = \dfrac{p(\boldsymbol{\theta}_i^{(t)})}{\sum_j w_j^{(t-1)} K_t(\boldsymbol{\theta}_i^{(t)}|\boldsymbol{\theta}_j^{(t-1)})}$.
 | | **end**
 | | Normalize $w_i^{(t)}$ over $i = 1, \ldots, P$.
 | **end**

end

4.2.2 Configurations of The Algorithm

There are several user-specified aspects and algorithm settings that have a major influence on the performance and accuracy of the algorithm. Thus, the configuration need to be chosen carefully, but for most of them no general rule of thumb exists. In the following, we discuss the different tuning parameters and settings, their influence on the performance and accuracy of the algorithm, and possible choices of the parameters.

Summary Statistics

The efficiency of the algorithm crucially depends on the marginal probability to observe the data set. Low probabilities of $p(\mathcal{D}^{\mathrm{obs}})$ have the consequence that the acceptance rate of the parameters is also low. Therefore, the calculation of the distances between the data sets $d(\mathcal{D}^{\mathrm{obs}}, \mathcal{D}^{\mathrm{sim}})$ is often replaced by the calculation of the distances $d(\mathcal{S}(\mathcal{D}^{\mathrm{obs}}), \mathcal{S}(\mathcal{D}^{\mathrm{sim}}))$ between summaries. As $p(\mathcal{S}(\mathcal{D}^{\mathrm{obs}}))$ is usually larger than $p(\mathcal{D}^{\mathrm{obs}})$, this results in a higher acceptance rate. If it holds for the summary statistics that $p(\boldsymbol{\theta}|\mathcal{D}) = p(\boldsymbol{\theta}|\mathcal{S}(\mathcal{D}))$, meaning that the statistic is sufficient, the true posterior can be obtained for $\epsilon \to 0$ (Nunes & Balding, 2010).

Examples for summary statistics are e.g. moments (see e.g. (Wegmann *et al.*, 2009; Beaumont *et al.*, 2002)) of the data. Other approaches that are based on concepts of hypothesis testing are proposed by Lillacci & Khammash (2013) and Ratmann *et al.* (2013). Given a predefined set of summary statistics, methods to choose summary statistics exist in the literature (see (Blum *et al.*, 2013) for a detailed comparison of methods to select summary statistics). As they mostly need a candidate set of statistics, this has the consequence that the efficiency of these methods depends on this initial choice of summary statistics.

Using summary statistics has some drawbacks. If the statistic does not capture enough information about the data, i.e. if its not sufficient, the difference between approximated and true posterior distribution is unknown (Marjoram & Tavaré, 2006). If the likelihood is unknown, it is not possible to identify whether a summary statistic is sufficient (Wilkinson, 2013).

Stopping Criterion

The decision whether the algorithm has converged or the approximation is good enough can be difficult. If the final tolerance level is too high, the approximation can be quite different from the true posterior. Conversely, if the tolerance level is too low, the computation time is longer than needed to obtain a reasonable approximation.

One possibility is to determine the minimal distance between observed and simulated data ϵ_{end} data-driven for example by subsampling. Assume there are n observed samples, then the following procedure can be repeated several times

- Randomly pick two equally-sized subsets $\mathcal{D}_1, \mathcal{D}_2 \subset \mathcal{D}^{obs}$ (e.g. with replacement)

- Calculate $\epsilon = d(\mathcal{S}(\mathcal{D}_1), \mathcal{S}(\mathcal{D}_2))$

and e.g. a certain percentile of the calculated ϵ values is chosen as final distance that needs to be achieved. This has the disadvantage that if the distance function depends on the number of samples n, the determined value may not represent the true minimal distance between observed and simulated data. Lenormand et al. (2013) presented an adaptive population Monte Carlo ABC scheme which additionally gives a stopping criterion based on the acceptance rate. But still, a minimal value for the acceptance rate needs to be defined.

Threshold Schedule

The choice of the ϵ-schedule is crucial for the performance of the algorithm. If the sequence decreases too slowly the computation time will be quite long until a reasonable approximation is achieved. On the other hand if it decreases too fast the acceptance rate will be too low which also results in a high computation time. The tolerances should always be determined considering the data, model and prior (Silk et al., 2013).

The most common approach is a quantile selection scheme. This approach determines ϵ_t at the beginning of the sampling of population t by calculating ϵ_t in a way that a given percentage of the particles in population $t - 1$ generated a smaller distance (Drovandi & Pettitt, 2011; Del Moral et al., 2012; Lenormand et al., 2013; Beaumont et al., 2009).

Silk *et al.* (2013) proposed to choose the next threshold based on an estimation of the acceptance rate curve.

Perturbation Kernel

The particles that are sampled for the intermediate and final distributions are additionally perturbed with respect to some perturbation kernel $K_t(\cdot|\cdot)$. For large numbers of parameters that need to be estimated from the data, or computational expensive models, the acceptance rate can drastically decrease if no suitable perturbation kernel is used (Filippi *et al.*, 2013). The trade-off between the acceptance rate, which tends to be better if the particles are not moved a lot, and the exploration of the parameter space, which needs a higher perturbation of the particle, has to be considered.

Filippi *et al.* (2013) listed several perturbation kernels, which aim to resemble the true posterior distribution. This list comprises e.g. component-wise kernels such as a uniform kernel and multivariate kernels such as a multivariate normal kernel. The former uses the width of the previous population that already fulfill the tolerance criterion for the current population to calculate the width of the kernel. The latter is based on the covariance of the particles of the previous population. A quite flexible and more local kernel is a multivariate normal kernel with k-nearest neighbors. This kernel uses a multivariate Gaussian distribution centered around the sampled parameter. The covariance is given by the empirical covariance of the k-nearest neighbors of the parameter, e.g. the k parameters that have the smallest euclidean distance to the sampled parameter.

Number of Particles Per Population

One advantage of ABC SMC is that it is able to obtained better approximations of multimodal posteriors than ABC rejection (Toni *et al.*, 2009). For this, it is necessary to find an appropriate number of particles P. The approximation gets better if more samples are used. But especially for the last populations the acceptance rates can be really low, which potentially results in a long computation time until the desired number of samples is obtained. Mostly, the number is predetermined by intuition and increased if the results are not reproducible.

Number of Samples to Generate for The Simulated Data Sets

Another parameter that needs to be chosen is the number of samples in the simulated data set denoted by m. Since the main idea is to reproduce the observed data set, an intuitive choice for m is to set it equal to the number of observed data points n. For large values of n this can result in long simulation times. Therefore, Lillacci & Khammash (2013) calculate within the method INSIGHT (see Section 4.2.3) the number of samples needed to decide whether to accept or reject an particles at a confidence level α.

The main advantage of ABC methods is that they are applicable to any stochastic model for which a forward simulation is possible. However, the methods have a lot of tuning parameters and configurations that have a great influence on the accuracy and performance of the algorithm. This can be seen as a huge disadvantage of the method, especially, as for most of the configurations no general applicable approach exists.

4.2.3 State-of-The-Art: ABC for Stochastic Models of Single-Cell Snapshot Data

ABC methods have been successfully applied for the analysis of single-cell snapshot data collected e.g. using flow cytometry. The *Inference for Networks of Stochastic Interactions among Genes using High-Throughput data (INSIGHT)* algorithm has already been used for high-dimensional models (Lillacci & Khammash, 2013). Since cells are discarded after being measured in flow cytometry, the measurements at the n_t different time points are independent. The distance between observed and simulated data sets can be calculated in the ABC rejection step using the maximal Kolmogorov-Smirnov (KS) distance:

$$d_{\text{KS}}(\mathcal{D}^{\text{obs}}, \mathcal{D}^{\text{sim}}) := \max_k \| \hat{F}_{\mathbf{X}_k} - \hat{G}_{\mathbf{Y}_k} \|_\infty \,, \tag{4.1}$$

with $\mathcal{D}^{\text{obs}} = \{\mathbf{X}_k\}_{k=1}^{n_t}$, $\mathcal{D}^{\text{sim}} = \{\mathbf{Y}_k\}_{k=1}^{n_t}$, and $\hat{F}_{\mathbf{X}_k}, \hat{G}_{\mathbf{Y}_k}$ being the corresponding empirical cumulative distributions. Here, a sample $\mathbf{X}_k$ contains the fluorescence levels of the n single-cells for a time point that is indexed by k and each $\mathbf{Y}_k$ comprises m samples. INSIGHT achieves good results, benefiting from large sample sizes provided by flow cytometry, from using the two-sample Kolmogorov-Smirnov test to compare the data sets, and from exploiting relationships between configurations of the ABC algorithm and the test statistic. Given some threshold ϵ, the critical value of $m_{\text{KS}}^{(c)}(\alpha, \epsilon)$, needed to decide whether to accept or reject the particle at a confidence level α, can be calculated by investigating properties of the KS distance (see (Lillacci & Khammash, 2013, Supplementary Information)):

$$
m_{\text{KS}}^{(c)}(\alpha, \epsilon) = \left\lceil \frac{-\log(\frac{1-\sqrt{1-\alpha}}{2})}{2 \left(\epsilon - \sqrt{-\frac{1}{2n} \log(\frac{1-\sqrt{1-\alpha}}{2})} \right)^2} \right\rceil . \tag{4.2}
$$

Given a fixed value for m, the critical value $\epsilon_{\text{KS}}^{(c)}(\alpha, m)$ can be computed:

$$
\epsilon_{\text{KS}}^{(c)}(\alpha, m) = \sqrt{\left(-\frac{1}{2n} \log \left(\frac{1 - \sqrt{1 - \alpha}}{2} \right) \right)} + \sqrt{\left(-\frac{1}{2m} \log \left(\frac{1 - \sqrt{1 - \alpha}}{2} \right) \right)} . \tag{4.3}
$$

We will adapt the idea of using test statistics for the development of an ABC SMC method for single-cell time-lapse data, which we later will compare with INSIGHT.

4.3 Approximate Bayesian Computation with Multivariate Test Statistics

One of the problems arising when inferring parameters from single-cell time-lapse data is the dimensionality of the time-series (Problem 1). This requires to study, how ABC can be used to infer parameters based on multivariate data. In the following, we develop an ABC method for single-cell time-lapse data using hypothesis testing (Lillacci & Khammash, 2013; Ratmann et al., 2013).

4.3.1 Multivariate Test Statistics

The goal is to decide whether to accept or reject a parameter based on the observed data and the data set that has been generated with respect to the parameter. In the context of test statistics, this can be done using a two-sample test. If the test indicates that both underlying distributions are equal, the parameter is accepted as sample from the posterior approximation. For the case of data sets that comprise only one-dimensional samples, tests relying on the KS distance can be used as proceeded in INSIGHT (Lillacci & Khammash, 2013). Since we want to apply ABC SMC with test statistics to multivariate data, we need to find an appropriate multivariate test statistic for the two-sample problem (see (Gretton *et al.*, 2012, Section 3.3.3) for an overview of multivariate two-sample tests).

In the following we introduce the cross-match test, which compares two distributions based on distances between samples, and the maximum mean discrepancy, which calculates the largest difference in expectation over functions in the unit ball of a reproducing kernel Hilbert space.

Cross-Match Test

Rosenbaum (2005) presented an exact and distribution-free test to compare two multivariate distributions based on multivariate samples, which for example consist of continuous or discrete vectors of possibly infinite dimension. For the test, which is named cross-match test, a complete graph is defined, in which nodes correspond to samples and edge weights correspond to distances between the samples. To obtain the test statistic a minimum weight non-bipartite matching is performed, i.e., the set of edges is found under the condition that every node is incident to exactly one edge of this set and the sum of weights of these edges is minimal. The number of cross-matches A_1, i.e., the matched pairs that comprising one observed and one simulated sample, then indicates, whether all samples are drawn from the same distribution. The null distribution of A_1 is

$$\Pr(A_1 = a_1) = \frac{2^{a_1} \left(\frac{n+m}{2} \right)!}{\binom{n+m}{n} a_0! a_1! a_2!}, \tag{4.4}$$

with a_l being the number of matches with exactly l observed samples. For the case of $n + m$ being uneven see (Rosenbaum, 2005). Note that the distribution only depends on the numbers of observed and simulated samples.

We want to decide whether to reject or accept the null hypothesis, based on c obtained cross-matches. If the probability to observe c or less cross-matches under the null hypothesis is greater than the confidence level α, we cannot reject the hypothesis. If the probability is smaller, we conclude that the samples are drawn from different distributions. Given

$$\Pr(A_1 \leq c) = \sum_{i=0}^{c} \frac{2^i \frac{n+m}{2}!}{\binom{n+m}{n}(\frac{n+m}{2} - \frac{n+i}{2})! i! (\frac{n-i}{2})!} \leq \alpha, \tag{4.5}$$

we can fix m and α and find numerically the critical value $c_{CM}^{(c)}(\alpha, m)$, the minimum number of cross-matches, which has to be exceeded to reject the null hypothesis. Moreover, we can fix α and c to find the minimum number of samples $m_{CM}^{(c)}(\alpha, c)$ needed to decide whether $A_1 > c$.

We implemented the cross-match test in MATLAB. For this, we integrated a blossom V algorithm[1] Kolmogorov, 2009 to perform the minimum-weight non-bipartite matching, which requires $\mathcal{O}((n + m)^3)$ arithmetic operations. Moreover, we use the euclidean distance for the calculation of the distances between the nodes. The cross-match test is visualized in Figure 4.1.

The main advantage of the cross-match test is that it is distribution-free and exact, i.e., it does not make assumptions about the underlying distribution and the null distribution is known in closed form.

Maximum Mean Discrepancy

An alternative multivariate test statistic for the two-sample problem is based on the maximum mean discrepancy (MMD), which has first been introduced by Borgwardt *et al.* (2006)

$$\text{MMD}[\mathcal{F}, p, q] := \sup_{f \in \mathcal{F}} \left(\mathbb{E}_p[f(x)] - \mathbb{E}_q[f(y)] \right).$$

If the distributions p and q are equal, the MMD is zero. Moreover, $\mathcal{F}$ is a class of functions $f : \mathcal{X} \to \mathbb{R}$ that is chosen to be the unit ball in a universal reproducing kernel Hilbert space $\mathcal{H}$, to achieve a trade-off between over- and underfitting. If $\mathcal{F}$ comprises not enough functions, the MMD may not be able to detect differences between the distributions p and q. Contrarily, if the class is to powerful, for $p = q$ the MMD may be significantly greater than zero for finite sample sizes. Given samples $\mathbf{X} = (\mathbf{x}_1, \ldots, \mathbf{x}_n)$ and $\mathbf{Y} = (\mathbf{y}_1, \ldots, \mathbf{y}_m)$ of p and q, respectively, an empirical estimate of the MMD is

---

[1]Available at http://pub.ist.ac.at/~vnk/software.html.

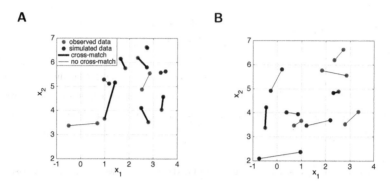

Figure 4.1: Illustration of the cross-match test. The blue dots mark the observed data $\mathcal{D}^{\mathrm{obs}}$, which has been generated by a bivariate normal distribution with co-variance matrix $\Sigma = \mathbf{I}_2$ and mean $\boldsymbol{\mu} = (2,5)^T$. The red dots mark the simulated data $\mathcal{D}^{\mathrm{sim}}$. The connected points are matched together by a minimum-weight non-bipartite matching. Bold lines highlight the cross-matches. (**A**) A higher number of cross-matches indicating a higher similarity of the data sets. The parameter would be accepted. (**B**) The particles produces a simulated data set, which is less similar to the observed data set in terms of number of cross-matches, and thus would be rejected.

$$\mathrm{MMD}[\mathcal{F}, \mathbf{X}, \mathbf{Y}] := \sup_{f \in \mathcal{F}} \left(\frac{1}{n} \sum_{i=1}^{n} f(\mathbf{x}_i) - \frac{1}{m} \sum_{j=1}^{m} f(\mathbf{y}_j) \right).$$

Using a kernel $k(\mathbf{x}, \mathbf{y}) = \Phi(\mathbf{x})^T \Phi(\mathbf{y})$ with nonlinear feature space mapping $\Phi(\mathbf{x})$ (see (Bishop, 2006) for further information), the MMD can be rewritten in terms of the mean embedding $\mu_p := E_p[\Phi(\mathbf{x})]$ as

$$\mathrm{MMD}[\mathcal{F}, p, q] = \sup_{f \in \mathcal{F}} \langle \mu_p - \mu_q, f \rangle = \| \mu_p - \mu_q \|_{\mathcal{H}}.$$

With $\mu_{\mathbf{X}} = \frac{1}{n} \sum_{i=1}^{n} \Phi(\mathbf{x}_i)$ and $k(\mathbf{x}, \mathbf{y}) = \langle \Phi(\mathbf{x}), \Phi(\mathbf{y}) \rangle$ the empirical estimate of MMD is

$$\mathrm{MMD}[\mathcal{F}, \mathbf{X}, \mathbf{Y}] = \left(\frac{1}{n^2} \sum_{i \neq j}^{n} k(\mathbf{x}_i, \mathbf{x}_j) + \frac{1}{m^2} \sum_{i \neq j}^{m} k(\mathbf{y}_i, \mathbf{y}_j) - \frac{2}{nm} \sum_{i,j=1}^{n,m} k(\mathbf{x}_i, \mathbf{y}_j) \right)^{\frac{1}{2}}. \quad (4.6)$$

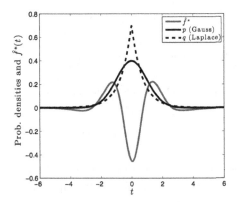

Figure 4.2: Function $\hat{f}^*$ witnesses the MMD between a Gaussian and Laplace distribution. This figure has been adopted from (Gretton *et al.*, 2012).

For the incorporation of the MMD into the ABC scheme, we exploit the following boundary (Gretton *et al.*, 2012, Corollary 9)

$$\text{MMD}[\mathcal{F}, \mathbf{X}, \mathbf{Y}] < \epsilon_{\text{MMD}}^{(c)}(\alpha, m) := \sqrt{2/m}\left(1 + \sqrt{2\log\alpha^{-1}}\right), \tag{4.7}$$

that assumes $m = n$. We use a MATLAB implementation of the MMD[2] that has been developed by Gretton *et al.* (2012). The computational costs for the evaluation of (4.6) are $\mathcal{O}((n + m)^2)$ and the test has shown to perform good even for low sample sizes and high dimensional data. A connection to summary statistics is given by the fact, that a feature map of a kernel is a sufficient statistic for the exponential family (Song, 2008).

4.3.2 Comparison of Test Statistics in ABC for Samples of a Bivariate Normal Random Variable

In the following, we assess the properties of the aforementioned multivariate test statistics and ABC SMC methods using them. We generate $n = 100$ samples $\mathbf{x}$ of a bivariate normal random variable with mean $\boldsymbol{\mu} = (\mu_1, \mu_2)^T = (0, 0)^T$ and the identity covariance matrix $\boldsymbol{\Sigma} = \mathbf{I}_2$ (see Figure 4.3A). We assume the covariance to be known, which yields two unknown parameters $\boldsymbol{\theta} = (\mu_1, \mu_2)^T$ that are estimated from the data. Using a multivariate normal conjugate prior with location parameter $\boldsymbol{\mu}_0$ and covariance $\boldsymbol{\Sigma}_0$ (Murphy, 2012)

---

[2]Available at http://www.gatsby.ucl.ac.uk/~gretton/mmd/mmd.htm.

$$p(\boldsymbol{\theta}) = \mathcal{N}(\boldsymbol{\theta}|\boldsymbol{\mu}_0, \boldsymbol{\Sigma}_0),$$

the posterior distribution is given by

$$p(\boldsymbol{\theta}|\mathcal{D}^{\mathrm{obs}}, \boldsymbol{\Sigma}) = \mathcal{N}(\boldsymbol{\theta}|\boldsymbol{\mu}_n, \boldsymbol{\Sigma}_n),$$
$$\boldsymbol{\Sigma}_n^{-1} = \boldsymbol{\Sigma}_0^{-1} + n\boldsymbol{\Sigma}^{-1}, \tag{4.8}$$
$$\boldsymbol{\mu}_n = \boldsymbol{\Sigma}_n\left(\boldsymbol{\Sigma}^{-1}(n\bar{x}) + \boldsymbol{\Sigma}_0^{-1}\boldsymbol{\mu}_0\right).$$

Here, $\bar{x}$ is the sample mean of the data. This posterior can be compared to the posterior approximation obtained by ABC SMC sampling. We implemented the ABC SMC Algorithm 4.2 in MATLAB, using a k-nearest neighbor perturbation kernel with $k = P/5$ and a 25th percentile approach for the selection of the threshold for the next population. The parameters for the prior are chosen as

$$\boldsymbol{\mu}_0 = \begin{pmatrix} 0 \\ 0 \end{pmatrix} \text{ and } \boldsymbol{\Sigma}_0 = \begin{pmatrix} 100 & 0 \\ 0 & 100 \end{pmatrix},$$

yielding the posterior parameters

$$\boldsymbol{\mu}_n = \begin{pmatrix} 0.0195 \\ -0.0202 \end{pmatrix} \text{ and } \boldsymbol{\Sigma}_n = \begin{pmatrix} 0.01 & 0 \\ 0 & 0.01 \end{pmatrix}. \tag{4.9}$$

Since the efficiency of the algorithm depends on configurations such as the threshold schedule, we only compare the approaches in terms of convergence, i.e., whether it is possible to obtain a reasonable approximation, and not in terms of performance.

Fixed Number of Simulations and Final Threshold Calculated by Test Statistic Inequalities

As there exists no general stopping criterion, we study, whether a final threshold can be determined by exploiting the relationship between distance value and number of simulations. We calculate the desired threshold of the last population corresponding to the given number of simulations $m = n$ according to (4.3), (4.5) and (4.7) and compare:

- Cross-match test (CM) with $c_{end} = c_{CM}^{(c)}(0.05, 100) = 40$.
- Maximum mean discrepancy (MMD) with $\epsilon_{end} = \epsilon_{MMD}^{(c)}(0.05, 100) = 0.4876$.
- Kolmogorov-Smirnov distance (KS) with $\epsilon_{end} = \epsilon_{KS}^{(c)}(0.05, 100) = 0.2956$.

Note that in contrast to MMD and KS, a high value for the CM indicates a good agreement. The resulting posterior approximations are shown in Figure 4.3B. The approximations of the posterior are much wider than the true posterior, which has been calculated using (4.8) and (4.9). This means that for this simple example, the critical threshold values can not be used as stopping criterion. We expect this approach to give an even worse threshold for higher dimensions and more complex examples.

Fixed Number of Simulations and Lower Final Tolerance

To study, whether a reasonable approximation of the posterior can be obtained with the distance calculations, we increased the desired number of cross-matches ($c_{end} = 56$) and decreased the final distances ($\epsilon_{MMD,end} = 0.0664, \epsilon_{KS,end} = 0.11$). We chose these values according to the 10th percentile of the distances obtained by simulating 10000 data sets with the true parameters and calculating the corresponding test statistics. Figure 4.3C shows the distributions of the test statistics. The gray shaded area indicates for which values of the statistic a parameter is accepted in the final population of ABC SMC. For CM, additionally the exact null distribution is visualized calculated with (4.4). We repeated the ABC SMC sampling for all distances, again for a fixed number of simulated samples $m = n$. The results are shown in Figure 4.3D. ABC with MMD and KS give a reasonable approximation of the posterior, while the posterior using the cross-match test is much wider than the true posterior distribution. Decreasing the tolerance did not improve the approximation significantly.

Adaptive Number of Simulations and Corresponding Threshold Schedule Using The KS Distance

The inequality for the KS distance can also be used if n and m are different. Therefore, we want to study, whether the method benefits from using an adaptive determination of m given some pre-defined threshold schedule, as proceeded in INSIGHT. We fixed the threshold schedule to $\epsilon = 0.99$, 0.7, 0.5, 0.4, 0.2 and 0.17 and calculated the corresponding critical values of $m = 4, 8, 18, 35, 802$, and 4435 using (4.2). The value of m for the last population in the adaptive approach is much higher than for $m = n$. Moreover, using the adaptive selection of m yields a total number of simulations in the ABC SMC algorithm which is more than twice as much as for $m = n$. The approximation obtained for this threshold value is much wider than the true posterior (see Figure 4.3D). An even lower final threshold ϵ_{end} increases the number of simulations for the last population drastically, since the value approaches the pol of (4.2) at $\epsilon = \sqrt{-\frac{1}{2n} \log \left(\frac{1 - \sqrt{1 - \alpha}}{2} \right)}$. Crossing this value yields a decreasing sequence for m which violates the assumptions of INSIGHT.

Inferring Parameters of The Covariance Matrix of a Bivariate Normal Distribution

We evaluate ABC SMC with MMD, CM and KS for a second data set comprising $n = 100$ samples (see Figure 4.4A). We generate samples of a bivariate normal distribution with known mean $\boldsymbol{\mu} = (0, 0)^T$ and a covariance matrix of the form

$$\Sigma = \begin{pmatrix} \theta_1 & \theta_2 \\ \theta_2 & \theta_1 \end{pmatrix}.$$

As the covariance matrix needs to be positive definite, both eigenvalues $\lambda_1 = \theta_1 + \theta_2$ and $\lambda_2 = \theta_1 - \theta_2$ need to be greater than zero. This yields the restrictions $\theta_1 + \theta_2 > 0$ and $\theta_1 - \theta_2 > 0$. Therefore, we use the following prior

$$p(\boldsymbol{\theta}) = \begin{cases} \frac{1}{100} & , \text{for } 0 < \theta_1 < 10, 0 \leq \theta_2 < \theta_1 \\ 0 & , \text{otherwise} \end{cases},$$

for the parameters $\boldsymbol{\theta} = (\theta_1, \theta_2)^T$, which are estimated in the following.

The final tolerances, $c_{\mathrm{end}} = 56$, $\epsilon_{\mathrm{MMD,end}} = 0.055$ and $\epsilon_{\mathrm{KS,end}} = 0.99$ are chosen with respect to the distances obtained by simulating data and calculating the statistics with the ground truth. The results of ABC SMC are visualized in Figure 4.4B. ABC SMC with MMD and CM is able to estimate the parameters. The confidence obtained using MMD is much higher than for CM. The KS approach provides an estimation of θ_1 only. The posterior approximation for θ_2 is much wider and only restricted by the relationship $|\theta_2| \leq \theta_1$. The difference can be explained by the lack of information included in the marginal distributions that are examined using KS. Information about θ_1 can only be

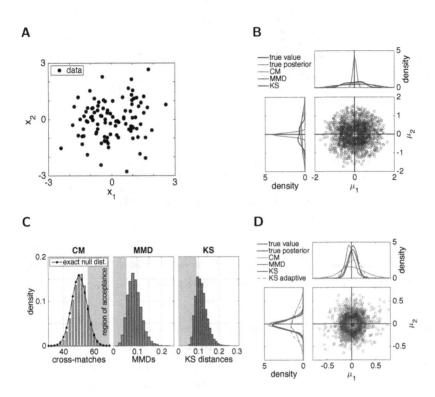

Figure 4.3: ABC SMC using test statistics to estimate the mean of a bivariate normal distribution. **(A)** Data generated from a bivariate normal distribution. **(B)** Posterior approximations for $m = n$ simulations for every population and final threshold calculated using inequalities for the test statistics. **(C)** Distributions of CM, MMD and KS statistics. The gray shaded area shows the region for which the parameters are accepted. **(D)** Posterior approximations with $m = n$ simulations for CM, MMD and KS with a lower tolerance to accept the corresponding parameters than in **(B)**. The dotted line shows the marginal distribution for the case of an adaptive number of simulations m with KS.

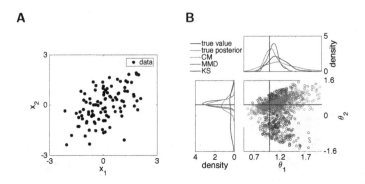

Figure 4.4: ABC SMC using test statistics to estimate entries of the covariance matrix of a bivariate normal distribution. (**A**) Depiction of 100 samples $\mathbf{x} \sim \mathcal{N}_2\left(\boldsymbol{\mu}, \boldsymbol{\Sigma}\right)$, with $\boldsymbol{\mu} = (0,0)^T$ and $\boldsymbol{\Sigma} = \left(\begin{smallmatrix} \theta_1 & \theta_2 \\ \theta_2 & \theta_1 \end{smallmatrix}\right)$, with $\theta_1 = 1$ and $\theta_2 = 0.5$. (**B**) Results of ABC with CM, MMD and KS. The yellow shaded are shows the region, where the prior $p(\boldsymbol{\theta}) > 0$.

gained by investigating the correlations among the measurements. The quality of the approximation did not improve significantly for lower tolerances.

We assessed ABC SMC with different approaches to incorporate CM, MMD and KS. For the estimation of the entries of the covariance matrix, only ABC SMC with multivariate statistics, CM and MMD, was able to estimate the parameters. The CM test requires a higher computation time than the MMD and yields less accurate approximations of the true posterior distribution for the example of a bivariate normal distribution. Accordingly, we use the MMD for our subsequent studies.

4.4 Simulation Example: Single-Cell Time-Series of a One-Stage Model of Gene Expression

In this section we apply the ABC SMC scheme described before to simulation examples of a one-stage model of gene expression (see Figure 4.5A).

4.4.1 Implementation

For the generation of artificial data of the one-stage model of gene expression we use a C implementation of the stochastic simulation algorithm (SSA) developed by Dennis Rickert. We implemented the ABC SMC algorithm in MATLAB according to Algorithm 4.2 and use the following settings:

- We sample from the posterior distribution of the $\log_{10}$-transformed parameters.

- For all scenarios, we use an uninformative component-wise uniform prior distribution $U[-6, 4]$ for the $\log_{10}$-transformed parameters.

- To compare simulated and observed data sets, we use a multivariate statistic, the MMD (4.6) and a univariate statistic, the maximal Kolmogorov-Smirnov distance (4.1).

- For the threshold schedule we use an adaptive quantile approach with the 25th percentile.

- We use a multivariate k-nearest neighbor perturbation kernel with $k = P/5$.

- We sample $P = 500$ particles per population.

- The final threshold ϵ_{end} is chosen in a data-driven fashion.

Since stochastic simulations can be computationally expensive for some proposed parameter combinations, we try to avoid a too high perturbation of the proposed particle, but still want to have enough flexibility to explore the parameters space. This is achieved by using a k-nearest neighbor perturbation kernel. We increase the number of particles and repeat the approximation if the result are not reproducible, i.e., if we do not obtain a similar approximation of the posterior within three repetitions. As we know the true parameters for the simulation study, we generate 1000 data sets using the true values of the parameters and calculate the corresponding distances. We use a threshold below the 5th percentile of those distances.

In the following, we consider two scenarios:

Scenario 1 The initial mRNA number is zero for all cells.

Scenario 2 The initial mRNA number is sampled from the steady state distribution.

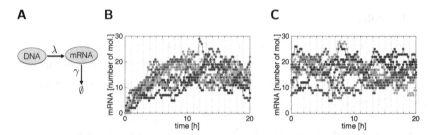

Figure 4.5: Illustration of artificial single-cell time-lapse data. (**A**) One stage model
of gene expression with mRNA synthesis rate λ and degradation rate γ.
(**B**) Artificial non-equilibrium time-series of $n = 10$ cells sampled every $\frac{1}{5}$ h.
(**C**) Equilibrium time-series of $n = 10$ cells measured at $n_t = 100$ time
points.

For both scenarios we generate $n = 10, 100$ and 1000 single-cell time-series for the synthesis
rate $\lambda = 5\,\mathrm{h}^{-1}$ and degradation rate $\gamma = 0.3\,\mathrm{h}^{-1}$ using the SSA. The initial conditions
are $[\mathrm{mRNA}](0) = 0$ for Scenario 1 and $[\mathrm{mRNA}](0) \sim \mathrm{Poi}(\lambda/\gamma)$ for Scenario 2 (Shahrezaei
& Swain, 2008). We simulate the system for $20\,\mathrm{h}$ and record the mRNA at $n_t = 100$
equidistant time points. The data sets are visualized for the case of $n = 10$ cells in
Figures 4.5B-C. For the evaluation of our method we assume λ and γ to be unknown and
estimate them from the data.

4.4.2 Out of Steady State Time-Series

In the following, we describe the results for Scenario 1, in which the population exhibits
transient behavior and the data comprises non-equilibrium time-series. Performing ABC
SMC with the settings described in Section 4.4.1, we obtain the posterior approximations
shown in Figures 4.6A and B. As expected, increasing the number of cells yields a more
narrow posterior distribution. The scatter plot of the samples shows that the parameters
are highly correlated. Moreover, the results are reproducible, indicated by the fact that
three repetitions of the sampling with the same parameter settings yield almost the same
approximation of the posterior distribution.

For the case of 100 cells and 100 measurements, the MAP estimates are given by $\boldsymbol{\theta}_{\mathrm{MMD}}^{\mathrm{MAP}} = (4.8259, 0.2945)^T$ and $\boldsymbol{\theta}_{\mathrm{KS}}^{\mathrm{MAP}} = (5.0069, 0.3078)^T$. We generated 1000 time-series based on
the MAP estimates and compared the mean and variance of the molecule numbers (see

Figure 4.6C) as well as mean and variance of the corresponding autocorrelation function (see Figure 4.6D). The fits and the corresponding properties of the data are almost indistinguishable.

We additionally compare the approximations with those obtained by the finite state projection (FSP) (Munsky & Khammash, 2006). This approach truncates the state space of the species and can be used for small and medium-sized systems. We sample from the posterior using a FSP-based likelihood and the MCMC toolbox DRAM (Haario et al., 2006). The results are shown in Figure 4.6. The approximations obtained by the ABC sampler are wider than the approximation using the FSP. ABC with MMD and KS yield similar results. To study the influence of the dimension of the time-series, we generated a scenario, in which 10 cells are measured at only 10 time points. While the approximation does not significantly improve compared to the FSP-based approximation for an increasing numbers of cells, the algorithm produces a better approximation for the case of measurements at $n_t = 10$ time points.

4.4.3 Steady State Time-Series

The results for Scenario 2 are shown in Figure 4.7. ABC SMC using KS is only able to estimate the fraction λ/γ (Figure 4.7B). This is explained by the fact that the marginals analyzed using the KS distance do not change over time. In contrast, the proposed multivariate methods using MMD exploits the temporal fluctuations and can infer both parameters (Figure 4.7A). The posterior distribution illustrated in Figure 4.7B differs only slightly from the initial uniform prior distribution. Surprisingly, the posterior distribution is shifted towards smaller parameter values and the true values are only at the boundary of the posterior approximations.

The fits for the case of $n = 100$ cells and $n_t = 100$ are shown in Figures 4.7C and D. The corresponding MAP estimates are $\boldsymbol{\theta}_{\text{MMD}}^{\text{MAP}} = (4.4388, 0.2650)^T$ and $\boldsymbol{\theta}_{\text{KS}}^{\text{MAP}} = (0.0005, 0.0003)^T$. As the kinetic rates estimated by the univariate consideration of the trajectories are much smaller than the true values, the corresponding number of molecules does not change within the simulation time of 20 h. Therefore, the autocorrelation function is not defined, since the corresponding time-series are constant.

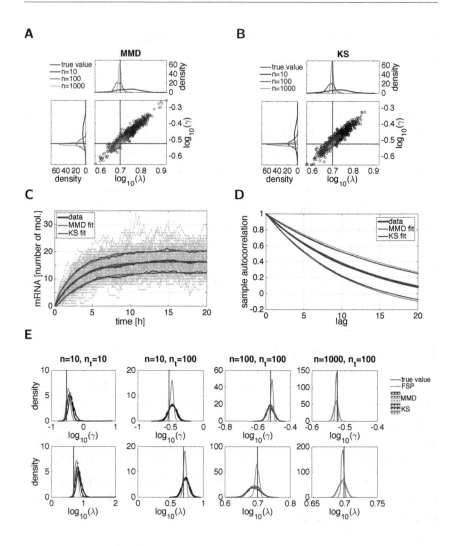

Figure 4.6: Results of ABC SMC for non-equilibrium time-series data. (**A, B**) Posterior
approximations obtained by ABC SMC with (**A**) MMD and (**B**) KS. (**C**)
Fitted mean and variance of molecule numbers for 1000 simulation generated
with the MAP estimates. Single trajectories are illustrated in gray. (**D**) Fitted mean and variance of the autocorrelation function for 1000 simulation
generated with the MAP estimates. The observed and simulated means and
variances are nearly indistinguishable for both, MMD and KS. (**E**) Comparison of marginals and FSP results for 10 cells and 10 time points (left),
and $10, 100, 1000$ cells with 100 measurements. The different lines show the
marginals for different repetitions of the sampling procedure.

Comparing the results with those obtained by the FSP (see Figure 4.7E) shows that ABC SMC with MMD produces a much wider approximation of the posterior distribution. The approximation is not as good as for Scenario 1, which can be explained by the fact that a lot of information about the parameters can be extracted by the MMD from the dynamics of the cells. The results for the case of less measurements ($n_t = 10$) are better than for the case of 100 measurements. This has also been observed for Scenario 1 (see Figure 4.6E).

4.4.4 Parameter Variability

Cell-to-cell variability of gene expression can be partitioned into intrinsic and extrinsic noise (Elowitz *et al.*, 2002). So far, only intrinsic noise has been considered, but the proposed approach can in principle also be used to infer extrinsic sources of cell-to-cell variability. In the following example, we model extrinsic noise by assuming variability in the mRNA synthesis and degradation rates. The parameters λ and γ are assumed to be log-normally distributed with means μ_λ, μ_γ and variances σ_λ^2, σ_γ^2. The data comprises $n = 100$ time-series measured at $n_t = 100$ time points. The true parameters used for the data generation are $\boldsymbol{\theta} = (\mu_\lambda, \sigma_\lambda^2, \mu_\gamma, \sigma_\gamma^2)^T = (5, 0.1, 0.3, 0.05)^T$. The time-series are depicted in Figure 4.8A. The overall variability is higher than in the scenarios without additional variability, e.g. as shown in Figure 4.5B and 4.6C.

The results obtained for the ABC SMC are depicted in Figure 4.8B. Here, also the intermediate distributions corresponding to different tolerance values are visualized showing the convergence of the algorithm. It reveals that the posterior distributions of the parameters μ_λ, μ_γ and σ_γ^2 are narrow, indicating identifiability. The posterior distribution for σ_λ^2 is wider and merely an upper bound can be determined. Accordingly, our analysis shows that in principle stochastic and deterministic variability can be reconstructed from single-cell time-lapse data.

4.4.5 Tree Structure

Single-cell time-lapse data often contain information about the ancestors of a cell (Etzrodt *et al.*, 2014). Using this information, ABC has e.g. been used to infer parameters by comparing differentiation probabilities (Marr *et al.*, 2012). We propose an approach to include tree-structured data in our ABC SMC sampler that uses the MMD to compare

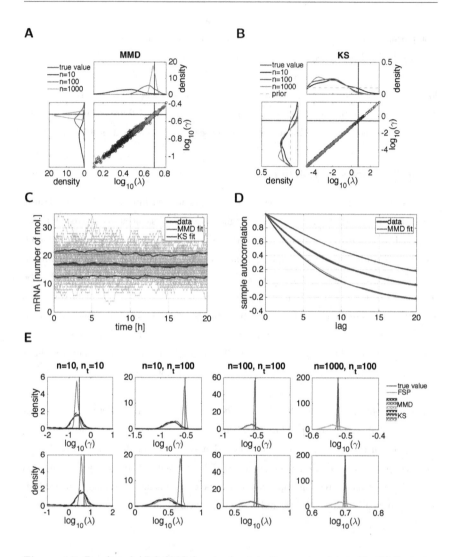

Figure 4.7: Results of ABC SMC for steady state time-series data. (**A**, **B**) Posterior
approximations obtained by ABC SMC with (**A**) MMD and (**B**) KS. (**C**)
Fitted mean and variance of number of molecules for 1000 simulation generated with the MAP estimates. The single trajectories are illustrated in
gray. (**D**) Fitted mean and variance of the autocorrelation function for 1000
simulation generated with the MAP estimates of the MMD. (**E**) Comparison
of marginals to results obtained by FSP for 10 cells and 10 time points (left),
and 10, 100, 1000 cells with 100 measurements. The different lines show the
marginals for different repetitions of the sampling procedure.

A   **B**

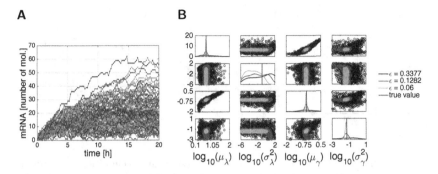

Figure 4.8: ABC SMC for cells affected by extrinsic noise. (**A**) Time-series of cells with parameter variability. (**B**) Results obtained by ABC SMC with MMD. On the diagonal, the marginal posterior distributions for the parameters are shown. The off diagonals provide scatter plots. The colours indicate the population corresponding to different tolerances ϵ and illustrate the convergence.

individual time-series. For this, we assume that a simple tree comprises one mother and its two daughter cells. One sample is given by $x_i = (x_{i,\text{mother}}, x_{i,\text{daughter}_1}, x_{i,\text{daughter}_2})$, as visualized in Figure 4.9A. Since the samples need to have the same dimension when using the MMD, we consider a fixed time interval before and after cell division. This is further motivated by the fact that the time-series exhibit transient dynamics after division before reaching the equilibrium, and therefore have a higher information content. Time-series of different lengths could also be interpolated and scaled to the same interval. To assess the method, we generate $n = 50$ simple trees (Figure 4.9B) that each includes one division process. A cell, which is measured at 50 time points, divides after 10 h. The molecules are equally split among its daughter cells. Both daughters are simulated for 10 h and measured at 50 time points. In the sample vector, the time-series of the mother is followed by the time-series of the cell, which has the higher molecule content at the end of the simulation.

Figure 4.9C visualizes the posterior approximations for three repetitions of the ABC sampling with MMD. As the posterior distributions are quite narrow, the parameters can be estimated with high confidence. This demonstrates the applicability of our method to not only time-series, but also single-cell time-lapse data with additional tree structure.

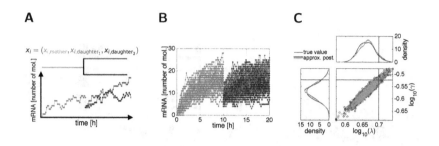

Figure 4.9: ABC SMC for tree structured data. (**A**) Sample comprising time-series of a mother and its two daughter cells. (**B**) Artifical data of $n = 50$ simple trees. (**C**) Posterior approximations for ABC SMC with MMD. The joint posterior as well as the marginal posteriors of the parameters λ and γ are shown for three repetitions of the sampling.

This approach, accounting for connections between the time-series of the mother and the daughter cells, might improve the insight gained into the underlying process. For example, parameters of the partitioning process could be estimated.

4.5 Discussion and Outlook

In this chapter, we introduced and evaluated an ABC SMC method to infer parameters of CTMCs. The method uses multivariate test statistics on the distribution of single-cell trajectories. We assessed our method for data from a bivariate normal distribution and for artificial single-cell time-series. We studied the use of different test statistics, comprising the multivariate statistics MMD and CM, and the univariate KS distance as used in INSIGHT (Lillacci & Khammash, 2013). For several examples, identifiability for the parameters was only achieved using multivariate statistics and ABC with MMD provided the best approximations of the posterior distributions. We found that for equilibrium single-cell time-lapse data the tracking information is important to identify the individual parameters.

A drawback of the method is the high computation time arising due to computationally expensive stochastic simulations. Thus, efficient simulation methods, such as τ-leaping (Gillespie, 2007) or the method of conditional moments (Hasenauer et al., 2014a) could be used instead of the SSA. These should be combined with appropriate threshold schedules (Silk *et al.*, 2013) and stopping criteria.

Using the acceptance region of the hypothesis test based on a given confidence level did not yield suitable approximations. This could arise due to low sample sizes, the higher dimension of the samples and imprecise boundaries for the test statistics. Thus it should be studied, whether better results can be achieved by considering more precise boundaries. It would probably also be worth to study how computation time can be saved by adapting the method to different numbers of observed and simulated samples ($m < n$). Furthermore, as more and more lineage information becomes available, its information content should be evaluated.

In summary, the proposed ABC SMC method using multivariate test statistics seems promising for the analysis of time-lapse single-cell data. It provides a general and flexible framework, which can easily be extended to similar data types. Using model selection, even sources of cell-to-cell variability might be unraveled.

5 Summary and Discussion

In this thesis, we studied models and parameter estimation methods for single-cell data that are able to account for cellular heterogeneity. We developed two methods, ODE-MMs with MEs for multivariate measurements (Chapter 3) and ABC SMC using multivariate test statistics (Chapter 4). While ODE-MMs have merely been used for single-cell snapshot data, the latter method, ABC SMC with multivariate statistics, can be applied to single-cell time-lapse data. Both of our methods can account for intrinsic and extrinsic noise sources and therefore are suitable tools for the elucidation of cell heterogeneity.

Two main extensions for ODE-MMs enable us on the one hand to account for cell-to-cell variability, and on the other hand to exploit correlations between multivariate measurements. Since the MEs that are used for the description of the subpopulation dynamics govern the evolution of the second order moments, the variances do not have to be estimated from the data. This yields a reduced number of parameters. Taking into account multivariate information of the data revealed a higher confidence in the estimates. In addition, we successfully tackled problems that arise when performing parameter estimation for advanced models. We proposed a numerically stable way to calculate the likelihood function of mixture probabilities and its gradient.

For the case of inferring parameters from single-cell time-lapse data, a key problem is the lack of a computational tractable likelihood function, which can be tackled by using likelihood-free ABC methods. We included temporal information of the data by considering single-cell trajectories as samples of a higher dimensional space. The distance between time-series needed for the ABC SMC algorithm is given by multivariate test statistics, which yields an improved parameter identifiability.

In conclusion, using more sophisticated modeling approaches rather than deterministic models enables us to extract more information from single-cell data. Since these models generally complicate parameter estimation, efficient and suitable methods are required. However, we presented two flexible methods for the analysis of single-cell snapshot and

single-cell time-lapse data. They consider both, appropriate models that account for
cell-to-cell variability and suitable approaches to calibrate those models to measurement
data. Therefore, applying our methods to experimental data could help to obtain a deeper
understanding of cell heterogeneity and the underlying biological processes.

Bibliography

Akaike, H. (1998). Information theory and an extension of the maximum likelihood principle. *Selected Papers of Hirotugu Akaike*. Springer, 199–213.

Altschuler, S. J. & Wu, L. F. (2010). Cellular heterogeneity: do differences make a difference?. *Cell* 141(4), 559–563.

Beaumont, M. A., Zhang, W. & Balding, D. J. (2002). Approximate Bayesian computation in population genetics. *Genetics* 162(4), 2025–2035.

Beaumont, M. A., Cornuet, J.-M., Marin, J.-M. & Robert, C. P. (2009). Adaptive approximate Bayesian computation. *Biometrika*, 1–8.

Bishop, C. M. (2006). *Pattern recognition and machine learning*. Vol. 4. 4. Springer New York.

Blum, M. G., Nunes, M. A., Prangle, D., Sisson, S. A., *et al.* (2013). A comparative review of dimension reduction methods in approximate Bayesian computation. *Statistical Science* 28(2), 189–208.

Borgwardt, K. M., Gretton, A., Rasch, M. J., Kriegel, H.-P., Schölkopf, B. & Smola, A. J. (2006). Integrating structured biological data by kernel maximum mean discrepancy. *Bioinformatics* 22(14), 49–57.

Burnham, K. P. & Anderson, D. R. (2002). *Model selection and multimodel inference: a practical information-theoretic approach*. Springer Science & Business Media.

Cho, K.-H. & Wolkenhauer, O. (2005). Systems Biology: Discovering the dynamic behaviour of biochemical networks. *BioSystems Review* 1(1), 9–17.

Csilléry, K., Blum, M. G., Gaggiotti, O. E. & François, O. (2010). Approximate Bayesian computation (ABC) in practice. *Trends in Ecology & Evolution* 25(7), 410–418.

Dargatz, C. (2010). Bayesian inference for diffusion processes with application in life sciences. PhD thesis. LMU Munich.

Davey, H. M. & Kell, D. B. (1996). Flow cytometry and cell sorting of heterogeneous microbial populations: the importance of single-cell analyses. *Microbiological Reviews* 60(4), 641–696.

Del Moral, P., Doucet, A. & Jasra, A. (2012). An adaptive sequential Monte Carlo method for approximate Bayesian computation. *Statistics and Computing* 22(5), 1009–1020.

Drovandi, C. C. & Pettitt, A. N. (2011). Estimation of parameters for macroparasite population evolution using approximate Bayesian computation. *Biometrics* 67(1), 225–233.

Elowitz, M. B., Levine, A. J., Siggia, E. D. & Swain, P. S. (2002). Stochastic gene expression in a single cell. *Science* 297(5584), 1183–1186.

Engblom, S. (2006). Computing the moments of high dimensional solutions of the master equation. *Applied Mathematics and Computation* 180(2), 498–515.

Etzrodt, M., Endele, M. & Schroeder, T. (2014). Quantitative Single-Cell Approaches to Stem Cell Research. *Cell Stem Cell* 15(5), 546–558.

Filippi, S., Barnes, C. P., Cornebise, J. & Stumpf, M. P. (2013). On optimality of kernels for approximate Bayesian computation using sequential Monte Carlo. *Statistical Applications in Genetics and Molecular Biology* 12(1), 87–107.

Gilks, W. R., Richardson, S. & Spiegelhalter, D. J. (1996). Introducing markov chain monte carlo. *Markov Chain Monte Carlo in Practice*.

Gillespie, D. T. (1977). Exact stochastic simulation of coupled chemical reactions. *The Journal of Physical Chemistry* 81(25), 2340–2361.

Gillespie, D. T. (1992). A rigorous derivation of the chemical master equation. *Physica A: Statistical Mechanics and its Applications* 188(1), 404–425.

Gillespie, D. T. (2007). Stochastic simulation of chemical kinetics. *Annual Review of Physical Chemistry* 58, 35–55.

Gretton, A., Borgwardt, K. M., Rasch, M. J., Schölkopf, B. & Smola, A. (2012). A kernel two-sample test. *The Journal of Machine Learning Research* 13(1), 723–773.

Haario, H., Laine, M., Mira, A. & Saksman, E. (2006). DRAM: efficient adaptive MCMC. *Statistics and Computing* 16(4), 339–354.

Hasenauer, J. (2013). Modeling and parameter estimation for heterogeneous cell populations. PhD thesis. Universität Stuttgart.

Hasenauer, J., Wolf, V., Kazeroonian, A. & Theis, F. J. (2014a). Method of conditional moments (MCM) for the Chemical Master Equation: a unified framework for the method of moments and hybrid stochastic-deterministic models. *Journal of Mathematical Biology* 69(3), 687–735.

Hasenauer, J., Hasenauer, C., Hucho, T. & Theis, F. J. (2014b). ODE Constrained Mixture Modelling: A Method for Unraveling Subpopulation Structures and Dynamics. *PLoS Computational Biology* 10(7), e1003686.

Hastie, T., Tibshirani, R. & Friedman, J. (2009). *The elements of statistical learning.* Vol. 2. 1. Springer.

Jahnke, T. & Huisinga, W. (2007). Solving the chemical master equation for monomolecular reaction systems analytically. *Journal of Mathematical Biology* 54(1), 1–26.

Kampen, N. van (2007). *Stochastic processes in physics and chemistry.* Amsterdam: North Holland, 3rd revised edition ed.

Kazeroonian, A., Fröhlich, F., Raue, A., Theis, F. J. & Hasenauer, J. (2016). CERENA: ChEmical REaction Network Analyzer - a toolbox for the simulation and analysis of stochastic chemical kinetics. *PLoS One (accepted).*

Kitano, H. (2002). Computational systems biology. *Nature* 420(6912), 206–210.

Kolmogorov, V. (2009). Blossom V: a new implementation of a minimum cost perfect matching algorithm. *Mathematical Programming Computation* 1(1), 43–67.

Lee, C. H., Kim, K.-H. & Kim, P. (2009). A moment closure method for stochastic reaction networks. *The Journal of Chemical Physics* 130(13), 134107.

Lenormand, M., Jabot, F. & Deffuant, G. (2013). Adaptive approximate Bayesian computation for complex models. *Computational Statistics* 28(6), 2777–2796.

Lillacci, G. & Khammash, M. (2010). Parameter estimation and model selection in computational biology. *PLoS Computational Biology* 6(3), e1000696.

Lillacci, G. & Khammash, M. (2013). The signal within the noise: efficient inference of stochastic gene regulation models using fluorescence histograms and stochastic simulations. *Bioinformatics*, 2311–2319.

Malone, J. H. & Oliver, B. (2011). Microarrays, deep sequencing and the true measure of the transcriptome. *BMC Biology* 9(1), 34.

Marin, J.-M., Pudlo, P., Robert, C. P. & Ryder, R. J. (2012). Approximate Bayesian computational methods. *Statistics and Computing* 22(6), 1167–1180.

Marjoram, P. & Tavaré, S. (2006). Modern computational approaches for analysing molecular genetic variation data. *Nature Reviews Genetics* 7(10), 759–770.

Marjoram, P., Molitor, J., Plagnol, V. & Tavaré, S. (2003). Markov chain Monte Carlo without likelihoods. *Proceedings of the National Academy of Sciences* 100(26), 15324–15328.

Marr, C., Strasser, M., Schwarzfischer, M., Schroeder, T. & Theis, F. J. (2012). Multiscale modeling of GMP differentiation based on single-cell genealogies. *FEBS Journal* 279(18), 3488–3500.

Michor, F. & Polyak, K. (2010). The origins and implications of intratumor heterogeneity. *Cancer Prevention Research* 3(11), 1361–1364.

Miyashiro, T & Goulian, M (2007). Single-cell analysis of gene expression by fluorescence microscopy. *Methods in Enzymology* 423, 458–475.

Munsky, B. & Khammash, M. (2006). The finite state projection algorithm for the solution of the chemical master equation. *The Journal of Chemical Physics* 124(4), 044104.

Munsky, B., Trinh, B. & Khammash, M. (2009). Listening to the noise: random fluctuations reveal gene network parameters. *Molecular Systems Biology* 5(318).

Murphy, K. P. (2012). *Machine learning: a probabilistic perspective*. MIT press.

Muzzey, D. & Oudenaarden, A. van (2009). Quantitative time-lapse fluorescence microscopy in single cells. *Annual Review of Cell and Developmental Biology* 25, 301–327.

Nunes, M. A. & Balding, D. J. (2010). On optimal selection of summary statistics for approximate Bayesian computation. *Statistical Applications in Genetics and Molecular Biology* 9(1).

Pritchard, J. K., Seielstad, M. T., Perez-Lezaun, A. & Feldman, M. W. (1999). Population growth of human Y chromosomes: a study of Y chromosome microsatellites.. *Molecular Biology and Evolution* 16(12), 1791–1798.

Pyne, S., Hu, X., Wang, K., Rossin, E., Lin, T.-I., Maier, L. M., Baecher-Allan, C., McLachlan, G. J., Tamayo, P., Hafler, D. A., *et al.* (2009). Automated high-dimensional flow cytometric data analysis. *Proceedings of the National Academy of Sciences* 106(21), 8519–8524.

Raftery, A. E. (1999). Bayes factors and BIC. *Sociological Methods & Research* 27(3), 411–417.

Ratmann, O., Camacho, A., Meijer, A. & Donker, G. (2013). Statistical modelling of summary values leads to accurate Approximate Bayesian Computations. *arXiv preprint arXiv:1305.4283.*

Raue, A., Kreutz, C., Maiwald, T., Bachmann, J., Schilling, M., Klingmüller, U. & Timmer, J. (2009). Structural and practical identifiability analysis of partially observed dynamical models by exploiting the profile likelihood. *Bioinformatics* 25(15), 1923–1929.

Raue, A., Schilling, M., Bachmann, J., Matteson, A., Schelke, M., Kaschek, D., Hug, S., Kreutz, C., Harms, B. D., Theis, F. J., *et al.* (2013). Lessons learned from quantitative dynamical modeling in systems biology. *PLoS ONE* 8(9), e74335.

Renart, J., Reiser, J. & Stark, G. R. (1979). Transfer of proteins from gels to diazobenzyloxymethyl-paper and detection with antisera: a method for studying antibody specificity and antigen structure. *Proceedings of the National Academy of Sciences* 76(7), 3116–3120.

Resat, H., Petzold, L. & Pettigrew, M. F. (2009). Kinetic modeling of biological systems. *Computational Systems Biology.* Springer, 311–335.

Rosenbaum, P. R. (2005). An exact distribution-free test comparing two multivariate distributions based on adjacency. *Journal of the Royal Statistical Society: Series B (Statistical Methodology)* 67(4), 515–530.

Schroeder, T. (2011). Long-term single-cell imaging of mammalian stem cells. *Nature Methods* 8(4), 30–35.

Schwarz, G. *et al.* (1978). Estimating the dimension of a model. *The Annals of Statistics* 6(2), 461–464.

Sengupta, B, Friston, K. & Penny, W. (2014). Efficient gradient computation for dynamical models. *NeuroImage* 98, 521–527.

Shahrezaei, V. & Swain, P. S. (2008). Analytical distributions for stochastic gene expression. *Proceedings of the National Academy of Sciences* 105(45), 17256–17261.

Silk, D., Filippi, S. & Stumpf, M. P. (2013). Optimizing threshold-schedules for sequential approximate Bayesian computation: applications to molecular systems. *Statistical Applications in Genetics and Molecular Biology* 12(5), 603–618.

Song, L. (2008). Learning via Hilbert space embedding of distributions. PhD thesis. University of Sydney.

Szekely, T. & Burrage, K. (2014). Stochastic simulation in systems biology. *Computational and Structural Biotechnology Journal* 12(20), 14–25.

Toni, T., Welch, D., Strelkowa, N., Ipsen, A. & Stumpf, M. P. (2009). Approximate Bayesian computation scheme for parameter inference and model selection in dynamical systems. *Journal of the Royal Society Interface* 6(31), 187–202.

Torres-Padilla, M.-E. & Chambers, I. (2014). Transcription factor heterogeneity in pluripotent stem cells: a stochastic advantage. *Development* 141(11), 2173–2181.

Wegmann, D., Leuenberger, C. & Excoffier, L. (2009). Efficient approximate Bayesian computation coupled with Markov chain Monte Carlo without likelihood. *Genetics* 182(4), 1207–1218.

Wilkinson, D. J. (2009). Stochastic modelling for quantitative description of heterogeneous biological systems. *Nature Reviews Genetics* 10(2), 122–33.

Wilkinson, R. D. (2013). Approximate Bayesian computation (ABC) gives exact results under the assumption of model error. *Statistical Applications in Genetics and Molecular Biology* 12(2), 129–141.

Printed in the United States
By Bookmasters